105 Topics in Current Chemistry

Fortschritte der Chemischen Forschung

Managing Editor: F. L. Boschke

Organic Chemistry

With Contributions by
A. J. Ashe III, G. Consiglio, E. L. Eliel, P. Pino

With 99 Figures and 22 Tables

Springer-Verlag
Berlin Heidelberg GmbH 1982

This series presents critical reviews of the present position and future trends in modern chemical research. It is addressed to all research and industrial chemists who wish to keep abreast of advances in their subject.

As a rule, contributions are specially commissioned. The editors and publishers will, however, always be pleased to receive suggestions and supplementary information. Papers are accepted for "Topics in Current Chemistry" in English.

ISBN 978-3-662-15348-2   ISBN 978-3-540-39449-5 (eBook)
DOI 10.1007/978-3-540-39449-5

Library of Congress Cataloging in Publication Data. Main entry under title: Organic chemistry.
(Topics in current chemistry = Fortschritte der chemischen Forschung; 105)
Bibliography: p. Includes index.
Contents: Prostereoisomerism (prochirality) / E. L. Eliel – – Asymmetrie hydroformyla-tion / P. Pino, G. Consiglio – – The group & heterobenzenes – – arsabenzene, stibabenzene, and bismabenzene / A. J. Ashe III. 1. Chemistry, Organic – – Addresses, essays, lectures. I. Ashe, A. J. (Arthur James), 1940 —. II. Series. Topics in current chemistry; 105.
QDI.F58 vol. 105 [QD255] 540s [547] 82-5974 AACR2

© by Springer-Verlag Berlin Heidelberg 1982
Originally published by Springer-Verlag Berlin Heidelberg New York in 1982
Softcover reprint of the hardcover 1st edition 1982

Table of Contents

Prostereoisomerism (Prochirality)

Ernest L. Eliel

Department of Chemistry, University of North Carolina, Chapel Hill, NC 27514, USA

1 Introduction. Terminology

Chirality (handedness, from Greek *"cheir"* = hand) is the term used for objects, including molecules, which are not superposable with their mirror images. Molecules which display chirality, such as (S)-$(+)$-lactic acid (*1*, Fig. 1) are called chiral. Chirality is often associated with a chiral center (formerly called an "asymmetric atom"), such as the starred carbon atom in lactic acid (Fig. 1); but there are other elements that give rise to chirality: the chiral axis as in allenes (see below) or the chiral plane, as in certain substituted paracyclophanes. [1,2]

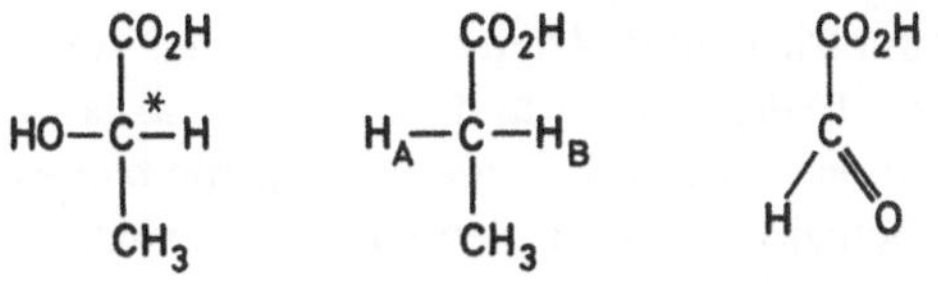

Fig. 1. Chiral and prochiral molecules

Often (e.g. in asymmetric synthesis) one is interested in the fact that in certain molecules, such as propionic acid (*2*, Fig. 1), an achiral center (here C_α) can be transformed into a chiral center by replacement of one or other of two apparently identical[1] ligands[2] by a different one. Thus the replacement of H_A at C_α in propionic acid (Fig. 1) by OH generates the chiral center of (S)-lactic acid whereas the analogous replacement of H_B gives rise to the enantiomeric (R)-lactic acid. C_α in propionic acid is therefore called a "prochiral center"[4]; H_A and H_B are called "heterotopic ligands" [5-7] (from Greek *"heteros"* = different and *"topos"* = place — see also below). Prochiral axes and planes may similarly be defined in relation to chiral axes and planes (see below).

Substitution is one of two common ways of interconverting organic molecules, the other is addition. The chiral center in lactic acid (Fig. 1) can, in principle, be generated by the addition of methylmagnesium iodide to the carbonyl group of glyoxylic acid (*3*, Fig. 1) (it might be necessary to protect the carboxylic acid group). Depending on which face of the aldehyde the Grignard reagent adds to, either (S)- or (R)-lactic acid is obtained. (The reader may convince himself that addition to the rear face of the aldehydic acid as depicted in Fig. 1 will give rise to (R)-lactic acid whereas (S)-lactic acid (*1*) is obtained by addition to the front face.) Thus the carbonyl group in glyoxylic acid is also said to be prochiral and to present two heterotopic faces.

Although the term prochirality is frequently used, especially by biochemists, it suffers from a limitation which arises from a corresponding limitation in the definition of chirality. Molecules may display purely stereochemical differences without being chiral: *cis-trans* isomers of olefins and certain achiral cis-trans isomers of cyclanes are examples. Thus (Fig. 2) (Z)- and (E)-1,2-dichloroethylene (*4*, *5*) are achiral diastereomers, as are *cis*- and *trans*-1,3-dibromocyclobutanes (*6*, *7*); being devoid of chirality these compounds have no chiral centers (or other chiral elements). Thus it is inappropriate to associate stereoisomerism with the occurrence of chiral

---

[1] The ligands must be identical when separated from the rest of the molecule. Such ligands have been called "homomorphic" (from Greek *"homos"* = same and *"morphe"* = form). [3]

[2] We use the term "ligand" to comprise both atoms and groups.

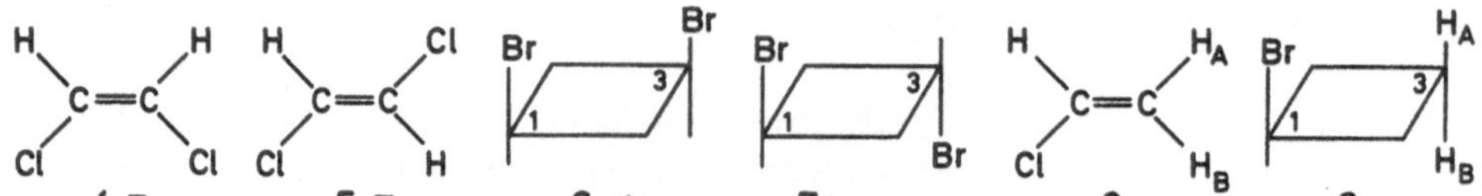

Fig. 2. Stereogenic and prostereogenic elements

elements; instead, we say that it is manifested by the existence of stereogenic elements [8] or elements of stereoisomerism [9]. These include stereogenic centers or centers of stereoisomerism (chiral centers, pseudoasymmetric centers [10] and centers of cis-trans isomerism in cyclanes, such as carbons No. 1 and 3 in 6 and 7, Fig. 2), axes of stereoisomerism or stereogenic axes (as in the olefins 4 and 5) and stereogenic planes or planes of stereoisomerism. Through the inclusion of these elements, cases such as the above (Fig. 2) of achiral stereoisomerism are properly taken into account.

Correspondingly, the concept of prochirality must be generalized to one of prostereoisomerism [3]. It is exemplified by chloroethylene (8, Fig. 2) and bromo-cyclobutane (9); these molecules display prostereoisomerism inasmuch as replacement of the homomorphic atoms H_A or H_B in 8 by chlorine gives rise to the stereoisomers 5 and 4, respectively. Similarly, replacement of H_A and H_B, respectively, in 9 by bromine gives rise to the stereoisomers 6 and 7. Thus 9 has a center of prostereoisomerism (or prostereogenic center) at C(3) and 8 has a prostereogenic axis (axis of prostereoisomerism) coinciding with the axis of the double bond. Again H_A and H_B in both 8 and 9 are heterotopic.

Cases of 'a prochiral axis (in allene 10, convertible by replacement of H_A by Cl into chiral allene 11 [10]) and a prochiral plane (in paracyclophane 12 which can be converted into the chiral structure 13 by replacement of H_A by CO_2H [12]) are shown in Fig. 3.

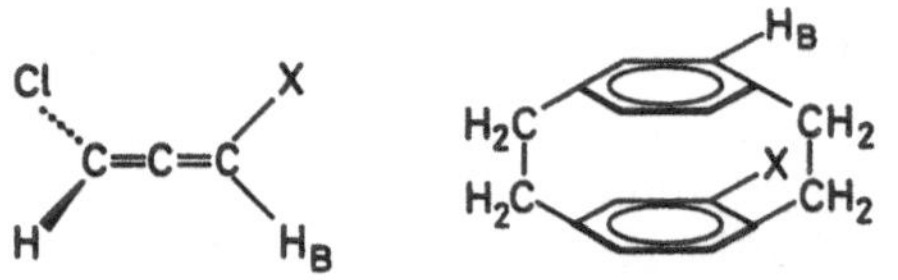

10 X = H_A
11 X = Cl

12 X = H_A
13 X = CO_2H

Fig. 3. Chiral and prochiral axes and planes

2 Significance. History

From the chemical point of view, the most significant aspect of the present subject lies in the possibility of differentiating the heterotopic ligands attached to elements of prostereoisomerism or the heterotopic faces of a prostereogenic double bond. The concept of heterotopic ligands and their recognition, in suitable instances, by NMR spectroscopy was first presented in a pioneering article by Mislow and Raban [5] (see also [25]) on which much of the subsequent discussion is based. Differentiation of heterotopic ligands or faces may be chemical or biochemical (as in asymmetric and stereoselective synthesis, including transformations by enzymes) or spectroscopic (notably by NMR spectroscopy). Before entering upon these topics in detail, we pose here a challenge to illustrate the utility of the concepts: In citric acid (*14*, Fig. 4), can the four methylene hydrogens H_A, H_B, H_C, H_D be distinguished by NMR spectroscopy, or by virtue of their involvement in the enzymatic dehydration of citric acid to *cis*-aconitic acid, or both? This question can easily be answered once the tenets of prostereoisomerism are understood: all the hydrogens can be distinguished by appropriate enzymatic reactions and H_A and H_B (as well as H_C and H_D) can give rise to distinct signals in the proton NMR spectrum whereas H_A and H_C (or H_B and H_D) give rise to coincident signals.

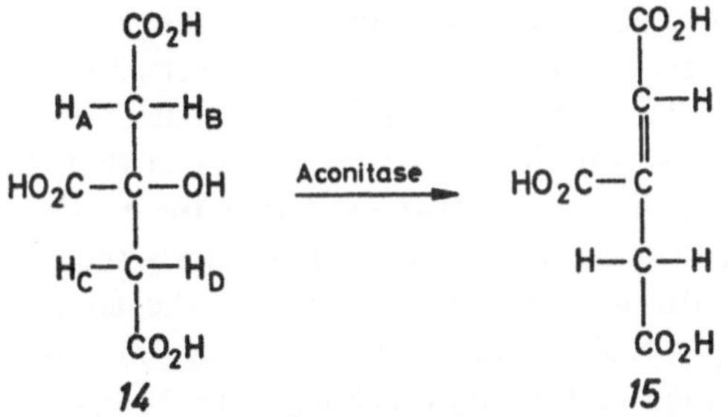

Fig. 4. Citric acid and aconitic acid

Historically, the first significant observation involving prochirality (though not recognized as such) was the decarboxylation of methylethylmalonic acid (*16*) to α-methylbutyric acid (*17*) in the presence of brucine [13] (Fig. 5). The C(2) carbon in α-methylbutyric acid is a chiral center; the C(2) carbon in the malonic acid precursor is a prochiral center. The product (Fig. 5) is optically active; indeed, this is one of the first recorded asymmetric syntheses. Clearly the superficial impression that the two carboxyl groups of the starting malonic acid are equivalent must be erroneous, for they can, in principle, be distinguished in the presence of the chiral catalyst brucine. A better-documented case is that of citric acid (Fig. 4). It was long known [14] that when oxaloacetic acid (*18*) labeled at C(4) is taken through the Krebs cycle, the α-ketoglutaric acid (*19*) formed is labeled exclusively at C(1) (next to the keto group), and not at all at C(5) (Fig. 6). This finding

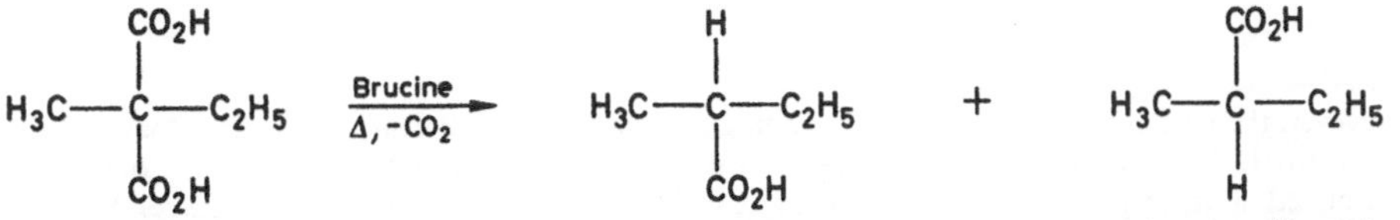

Fig. 5. Asymmetric decarboxylation of methylethylmalonic acid

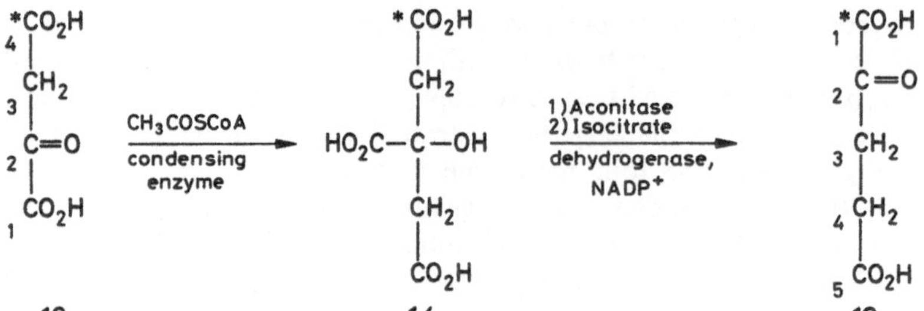

Fig. 6. Part of citric acid cycle

seemed to throw doubt on the theretofore assumed intermediacy of citric acid (*14*) in the cycle since, it was argued, the two ends of citric acid ($-CH_2CO_2H$) are equivalent and therefore the α-ketoglutaric acid formed through this intermediate should be labeled equally at C(1) and C(5). However, it is now clear (see Sects. 3 and 5) that the experiment in no way eliminates citric acid as a potential intermediate in the oxaloacetic acid — α-ketoglutaric acid transformation by virtue of the fact that the two CH_2CO_2H branches are, in fact, distinct and distinguishable by enzymes because of the prochiral nature of the central carbon atom, C(3). Similarly the fact that phosphorylation of glycerol (*20*) with ATP in the presence of the enzyme glycerokinase gives exclusively (*R*)-(—)-glycerol-1-phosphate [15] (*21*, Fig. 7) shows that the enzyme can distinguish between the two primary alcohol groups of glycerol and that these groups must thus be distinct: C(2) in glycerol is prochiral.

Fig. 7. Enzymatic phosphorylation of glycerol

The first[3] glimpse of understanding came when Ogston [16a] pointed out that an attachment of a substrate Caa'bc (a = a') to an enzyme at three sites (so-called "three-point contact") could lead to the observed distinction between the homomorphic (as we would now say) groups a and a', as shown in Fig. 8. If A is a catalytically active site on the enzyme and B and C are binding sites, Fig. 8 shows that only a but not a' can be brought into juxtaposition with the active site A

---

[3] However, the first mention of a three-point contact (between a chiral drug and its receptor) is found in an article by Easson and Stedman published in 1933 [16b], and a year later, Max Bergmann postulated a three-point contact (involving CO_2H, H_2N and the dipeptide linkage) between peptidases and the dipeptides hydrolyzed by them. [16c] Both these publications seem to have been overlooked subsequently; I thank Professors V. Prelog and H. Hirschmann, respectively, for drawing my attention to them.

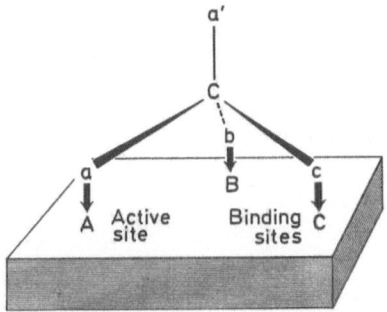

Fig. 8. Ogston's three-point contact model
[From H. Hirschmann, "Newer Aspects of Enzymatic Stereochemistry" in Comprehensive Biochemistry. Vol. 12, M. Florkin and E. H. Stotz, eds. By permission of Elsevier Publishing Co.]

when b and c are bound to B and C. Therefore a but not a' may be enzymatically transformed; a and a' are clearly distinguishable.

It was subsequently recognized that, whereas Ogston's picture provides a mechanistic rationale for the observed distinction of *apparently* equivalent groups (for a more detailed picture, see Fig. 56) it does not provide a unique rationale. Indeed, the distinguishability of the a and a' groups in Caa'bc (a prochiral center), is a consequence of symmetry properties and is independent of any mechanistic principles. This was probably first recognized by Schwartz and Carter [17] who called such carbon atoms (Caabc) "meso carbons" (now supplanted by "prochiral carbon atoms"). Excellent reviews concerning the nature of prochiral centers are now available [5, 18, 19].

Before entering upon the substance of the matter, we must pick up one other, at the beginning apparently unrelated, historical thread. In 1957 two groups of investigators [20, 21] discovered that in molecules of the type CX_2YCabc (for example $CF_2BrCHBrC_6H_5$ or $CH_2BrC(CH_3)BrCO_2CH_3$) the X nuclei (F in the first example, H in the second) displayed distinct NMR signals. Although the phenomenon was not clearly understood until some time later [22, 23], it is now clear that the non-equivalence of such X-nuclei in nuclear magnetic resonance rests on symmetry principles [5], as does the earlier-mentioned non-equivalence in enzymatic reactions and other reactions involving chiral reagents. The next threee sections (3–5) will deal with the explanation of these non-equivalencies and their chemical and spectral consequences.

3 Homotopic and Heterotopic Ligands and Faces [5, 7, 18]

3.1 Homotopic Ligands and Faces

We have indicated in the previous section that some apparently alike ligands are, in fact, not equivalent towards enzymes or in their NMR signals. How does one

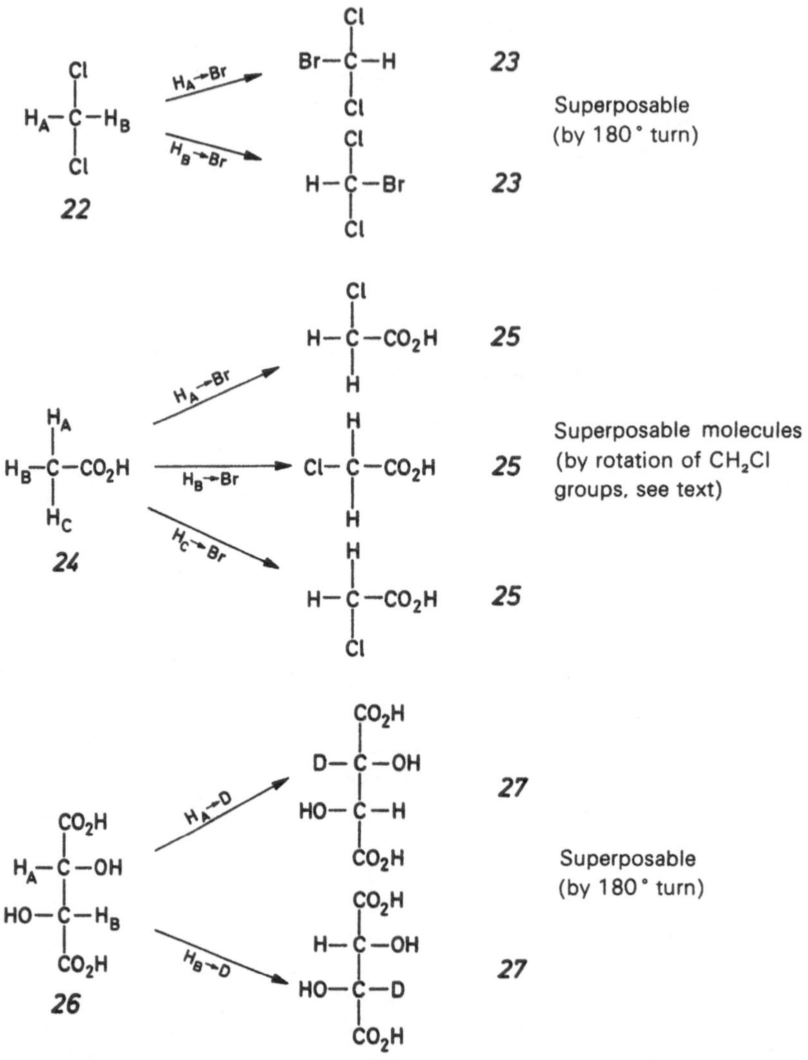

$$
\begin{array}{c}
\text{Cl} \\
| \\
\text{H}_A-\text{C}-\text{H}_B \\
| \\
\text{Cl} \\
\mathbf{22}
\end{array}
\quad
\xrightarrow{H_A \to Br}
\quad
\begin{array}{c}
\text{Cl} \\
| \\
\text{Br}-\text{C}-\text{H} \\
| \\
\text{Cl} \\
\end{array}
\quad \mathbf{23}
$$

$$
\xrightarrow{H_B \to Br}
\quad
\begin{array}{c}
\text{Cl} \\
| \\
\text{H}-\text{C}-\text{Br} \\
| \\
\text{Cl} \\
\end{array}
\quad \mathbf{23}
$$

Superposable
(by 180° turn)

$$
\begin{array}{c}
\text{H}_A \\
| \\
\text{H}_B-\text{C}-\text{CO}_2\text{H} \\
| \\
\text{H}_C \\
\mathbf{24}
\end{array}
$$

$$
\xrightarrow{H_A \to Br}
\quad
\begin{array}{c}
\text{Cl} \\
| \\
\text{H}-\text{C}-\text{CO}_2\text{H} \\
| \\
\text{H} \\
\end{array}
\quad \mathbf{25}
$$

$$
\xrightarrow{H_B \to Br}
\quad
\begin{array}{c}
\text{H} \\
| \\
\text{Cl}-\text{C}-\text{CO}_2\text{H} \\
| \\
\text{H} \\
\end{array}
\quad \mathbf{25}
$$

$$
\xrightarrow{H_C \to Br}
\quad
\begin{array}{c}
\text{H} \\
| \\
\text{H}-\text{C}-\text{CO}_2\text{H} \\
| \\
\text{Cl} \\
\end{array}
\quad \mathbf{25}
$$

Superposable molecules
(by rotation of CH_2Cl
groups, see text)

$$
\begin{array}{c}
\text{CO}_2\text{H} \\
| \\
\text{H}_A-\text{C}-\text{OH} \\
| \\
\text{HO}-\text{C}-\text{H}_B \\
| \\
\text{CO}_2\text{H} \\
\mathbf{26}
\end{array}
$$

$$
\xrightarrow{H_A \to D}
\quad
\begin{array}{c}
\text{CO}_2\text{H} \\
| \\
\text{D}-\text{C}-\text{OH} \\
| \\
\text{HO}-\text{C}-\text{H} \\
| \\
\text{CO}_2\text{H} \\
\end{array}
\quad \mathbf{27}
$$

$$
\xrightarrow{H_B \to D}
\quad
\begin{array}{c}
\text{CO}_2\text{H} \\
| \\
\text{H}-\text{C}-\text{OH} \\
| \\
\text{HO}-\text{C}-\text{D} \\
| \\
\text{CO}_2\text{H} \\
\end{array}
\quad \mathbf{27}
$$

Superposable
(by 180° turn)

Fig. 9. Homotopic ligands

decide, then, whether nuclei are equivalent?[4] There are two criteria, a substitution criterion and a symmetry criterion. Similar criteria (addition or symmetry) serve to test the equivalency of faces.

---

[4] The term "equivalent" is overly general and therefore bland and of equivocal meaning. Thus the methylene hydrogen atoms in propionic acid (Fig. 1) are equivalent when detached (i.e. they are homomorphic), but, as already explained, they are not equivalent in the $CH_3CH_2CO_2H$ molecules because of their placement — i.e. they are heterotopic. Ligands that are equivalent by the criteria to be described in the sequel are called "homotopic" from Greek "homos" = same and "topos" = place [6], those that are not are called "heterotopic".

3.1.1 Substitution and Addition Criteria [5]

Two homomorphic ligands (see footnotes on p. 3) are homotopic if replacement of first one and then the other by a different ligand[5] leads to the same structure. Thus, as shown in Fig. 9, the two hydrogen atoms in methylene chloride (*22*) are homotopic because replacement of either by, say, bromine gives the same $CHBrCl_2$ molecule (*23*); the three methyl hydrogens in acetic acid (*24*) are homotopic because replacement of any one of them by, say, chlorine gives one and the same chloroacetic acid (*25*); the two methine hydrogens in (*R*)-(+)-tartaric acid (*26*) are homotopic because replacement of either of them e.g. by deuterium gives the same (2*R*,3*R*)-tartaric-2-*d* acid (*27*).

Two corresponding faces of a molecule (usually but not invariably faces of a double bond) are homotopic when addition of the same reagent to either face gives the same product. For example, addition of HCN to acetone (*28*) will give the same cyanohydrin *29*, no matter to which face addition occurs (Fig. 10) and addition of bromine to ethylene similarly gives $BrCH_2CH_2Br$, regardless of the face of approach. The two faces of the $C=O$ double bond of acetone and of the $C=C$ double bond of ethylene are thus homotopic.

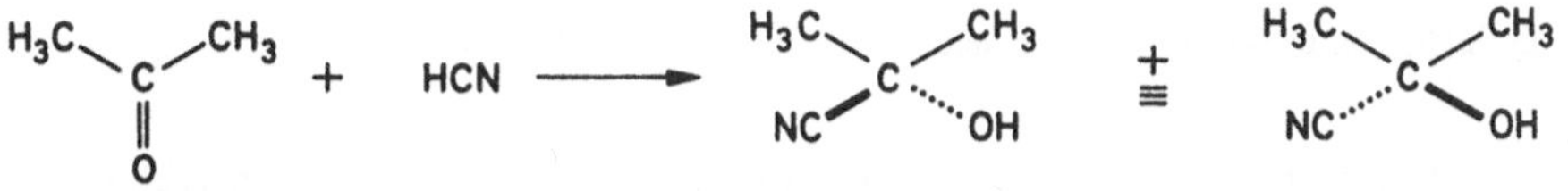

Fig. 10. Homotopic faces: Addition of HCN to acetone

3.1.2 Symmetry Criterion [5]

Ligands are homotopic if they can interchange places through operation of a C_n symmetry axis. Thus the chlorine atoms in methylene chloride (symmetry point group C_{2v}) are homotopic since they exchange places through a 180° turn around the C_2 axis (C_2^1). Similarly, the methine hydrogens of (+)-tartaric acid (Fig. 9) are interchanged by operation of the C_2 axis (the molecule belongs to point group C_2). The situation in acetic acid is somewhat more complicated. If we depict this molecule as stationary in one of its eclipsed conformations, we see (Fig. 11) that the hydrogens are heterotopic. However, rotation around the H_3C-CO_2H axis is rapid on the time scale of most experiments. We are therefore dealing with a case of averaged symmetry leading to interchange of the three methyl hydrogens of CH_3CO_2H, which are thus homotopic when rotation is fast on the time scale of whatever experiment is being considered.

---

[5] The replacement ligand must be different not only from the original one but also from all other ligands attached to the same atom. For example (Fig. 9) one cannot test the equivalency of the hydrogen atoms in methylene chloride by replacing them by chlorine or the equivalency of the hydrogens in (+)-tartaric acid or acetic acid by replacing them by CO_2H groups. The reason for this restriction will soon become obvious.

It is important to recognize that the presence of a symmetry axis in a molecule does not guarantee that homomorphic ligands will be homotopic. It is necessary that operation of the symmetry axis make the nuclei in question interchange places. Thus in 1,3-dioxolane (Fig. 12), in its average planar conformation, the hydrogens at C(2) are homotopic since they are interchanged by operation of the C_2 axis (the symmetry point group of the molecule is C_{2v}). On the other hand, the geminal hydrogen atoms at C(4) [or C(5)] are not interconverted by the C_2 symmetry operation and are therefore heterotopic (H_A with respect to H_B and H_C with respect to H_D). However H_A and H_D are homotopic (as are H_B and H_C), being interchanged once again by the C_2 axis.

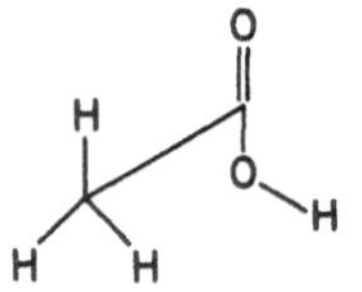

Fig. 11. Eclipsed conformation of acetic acid 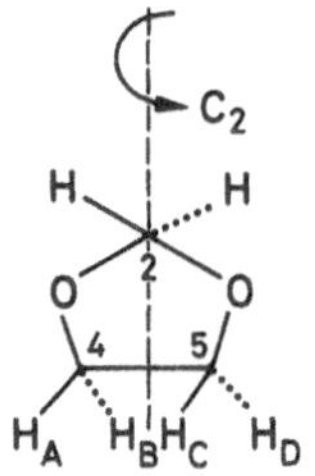   Fig. 12. 1,3-Dioxolane

Faces of double bonds are similarly homotopic when they can be interchanged by operation of a symmetry axis. (Since there are only two such faces, the pertinent axis must, of necessity, be of even multiplicity so as to contain C_2.) Thus the two faces of acetone (Fig. 10) are interchanged by the operation of the C_2 axis (the molecule is of symmetry C_{2v}); the two faces of ethylene (D_{2h}) are interchanged by operation of two of the three C_2 axes (either the one containing the C=C segment or the axis at right angles to the first one and in the plane of the double bond).

By way of an exercise, the reader may convince himself that the two hydrogens in each of the three dichloroethylenes (1,1-, cis-1,2-, trans-1,2-) and the four hydrogens in methane, CH_4, allene, $H_2C=C=CH_2$ and ethylene, $H_2C=CH_2$ are homotopic. It might be noted that, in a rigid molecule, the number of homotopic ligands in a set cannot be greater than the symmetry number of the molecule in question. Thus the four hydrogen atoms H_{A-D} in 1,3-dioxolane (Fig. 12) cannot possibly all be homotopic, since the symmetry number of the molecule is only 2. Similarly, rigid molecules in the nonaxial point groups C_1, C_s and C_i cannot display homotopic ligands because $\sigma = 1$ for these groups; the same is true of $C_{\infty v}$. (As mentioned above, this does not apply to cases of averaged symmetry such as acetic acid which has homotopic methyl hydrogens even though its non-averaged symmetry point group is C_s.)

Returning to the above-mentioned unsaturated compounds, the reader might also note that the faces in four of the five are homotopic, the exception being trans-1,2-dichloroethylene which has heterotopic faces.[6]

---

[6] The addition criterion tends to be confusing when applied to a molecule like ethylene where addition occurs at both ends of the double bond. The reader is advised, in such cases, either to use the symmetry criterion or to choose epoxidation as the test reaction for the addition criterion. For additional examples involving the heterotopic faces of not only olefins and carbonyl compounds

3.2 Enantiotopic Ligands and Faces

Just as one divides stereoisomers into two sets, enantiomers (Greek *enantios* = opposite) and diastereomers, so it is convenient to divide heterotopic (non-equivalent) groups or faces into enantiotopic and diastereotopic moieties. Enantiotopic ligands are ligands which find themselves in mirror-image positions whereas diastereotopic ligands are in stereochemically distinct positions not related in mirror-image fashion; similar considerations relate to planes of double bonds.

The two criteria used to spot homotopic ligands and faces may also be used to detect those which are enantiotopic.

3.2.1 Substitution-Addition Criterion

Two ligands are enantiotopic if replacement of either one of them by a different achiral ligand[7] (see also footnote 5 on p. 9) gives rise to enantiomeric products. Examples are shown in Fig. 13. The marked hydrogens (H_A, H_B) in CH_2ClBr (*30*), meso-tartaric acid (*32*), cyclobutanone (*34*) [at C(2) and C(4) but not C(3)] and chloroallene (*36*) [at C(3)] are enantiotopic, as are the methyl carbons in isopropyl alcohol (*38*). meso-Tartaric acid, incidentally, exemplifies one of the rare instances of a molecule with heterotopic ligands but no discernible prochiral center or other element of prochirality.

Those encountering the phenomenon of enantiotopic ligands for the first time are sometimes puzzled by the nature of the difference between such ligands. One way of explaining the difference is by the very substitution criterion: if replacement of two ligands, in turn, by a third one gives rise to different (enantiomeric) products, then the ligands can, *by definition*, not be equivalent (homotopic). A perhaps more satisfying view of the matter [17, 25] is shown in Fig. 14. If ones views the rest of the CH_AH_BClBr molecule from the vantage point of H_A one perceives the atoms Br—Cl—H_B in a counterclockwise direction (A). Contrariwise, if one views the remainder of the molecule from H_B, Br—Cl—H_A are seen in a clockwise sequence (B). Therefore the environment of H_B is the mirror image of the environment of H_A. We shall return to this view in Section 6.

Similar criteria, but of addition, can be established for enantiotopic faces. Faces are enantiotopic if addition of the same chiral reagent[7] to either one or the

---

but also species of the oxime or hydrazone type, $RR'C=N{\diagup}^X$, planar trigonal species, such as carbonium ions, $RR'R''C^+$, and even bent disubstituted atoms bearing unshared electron pairs, such as sulfides, $RR'S$: and related compounds, the reader should consult Ref. 24. In this reference, a double bond of an olefin is considered as a single entity whereas in Ref. 4 each end of the double bond is considered separately. This makes a difference, for example in *cis*-2-butene where the two faces of the double bond are homotopic overall, whereas the two faces of each $CH_3CH=$ moiety taken separately are heterotopic. This point will be discussed further below.

[7] If the test ligands are chiral, the products of replacing first one and then the other of two enantiotopic ligands by them will be diastereomeric. Similar considerations apply to addition of chiral ligands to enantiotopic faces.

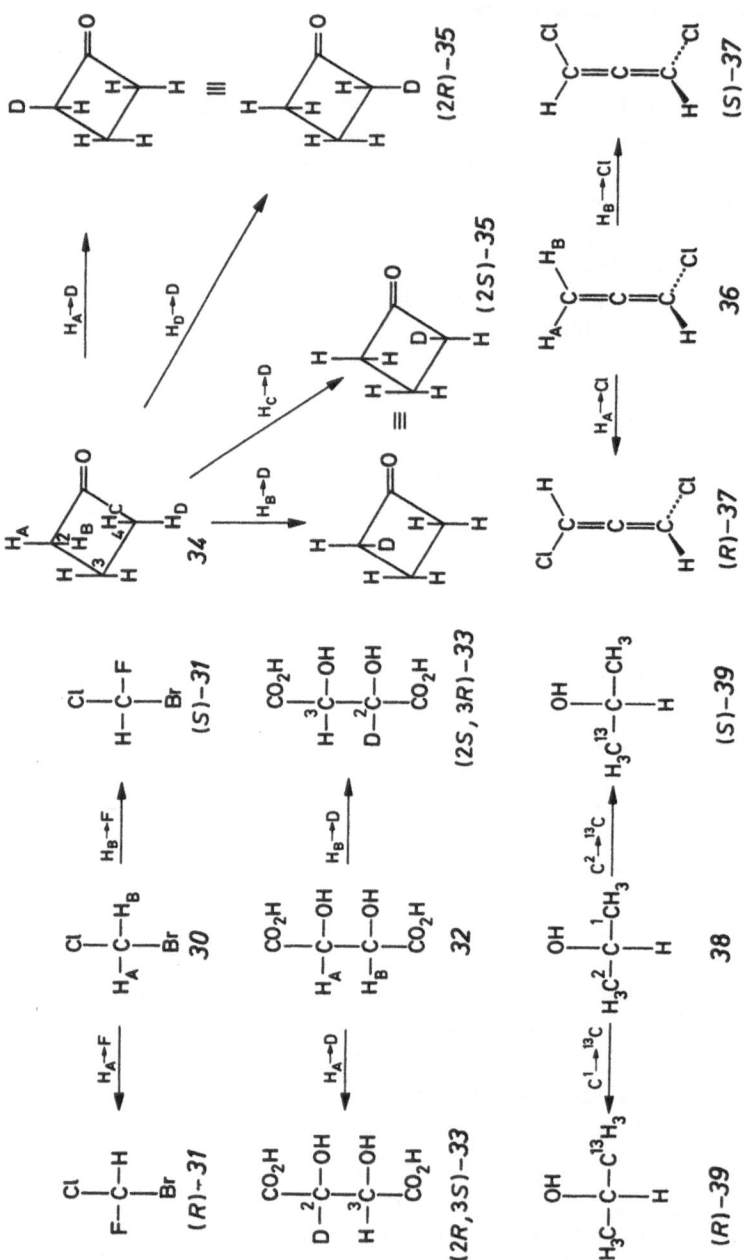

Fig. 13. Enantiotopic ligands

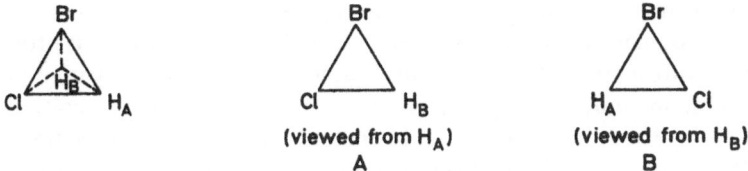

Fig. 14. CH₂ClBr — View of the rest of the molecule from each enantiotopic ligand

other will give rise to enantiomeric products. Thus (Fig. 15) addition of HCN to the two enantiotopic faces of acetaldehyde gives rise to the two enantiomers of lactonitrile. (Here, also — see footnote 5 on p. 8 — the added group must be different from any group already there. Thus we cannot test the enantiotopic nature of the two faces of the C=C function of acetaldehyde (*40*) by addition of CH₃MgI, since the group added — CH₃ — is the same as one of the existing groups.)

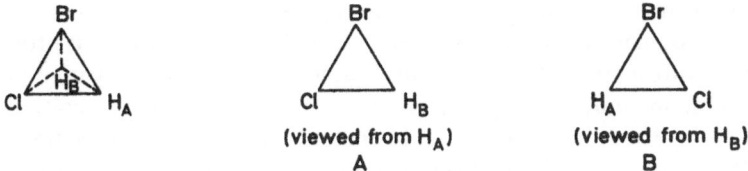

Fig. 15. Addition of HCN to acetaldehyde

3.2.2 Symmetry Criterion

Enantiotopic ligands and faces are not interchangeable by operation of a symmetry element of the first kind (C_n, simple axis of symmetry) but must be interchangeable by operation of a symmetry element of the second kind (σ, plane of symmetry; i, center of symmetry or S_n, alternating axis of symmetry). (It follows that, since chiral molecules cannot contain a symmetry element of the second kind, there can be no enantiotopic ligands or faces in chiral molecules. Nor, for different reasons, can such ligands or faces occur in linear molecules, $C_{\infty v}$ or $D_{\infty h}$.)

The symmetry planes (σ) in molecules *30, 32, 34, 36, 38,* Fig. 13 should be readily evident. It is possible to have both homotopic and enantiotopic ligands in the same set, as exemplified by the case of cyclobutanone (*34*): H_A and H_D are homotopic as are H_B and H_C. H_A is enantiotopic with H_B and H_C; H_D is similarly enantiotopic with H_C and H_B. The sets $H_{A,B}$ and $H_{C,D}$ may be called equivalent (or homotopic) sets of enantiotopic hydrogen atoms. The unlabeled hydrogens at position 3, constitutionally distinct — see Section 3.4 — from those at C(2, 4), are homotopic with respect to each other. Enantiotopic ligands need not be attached to the same atom — *viz.* the case of *meso*-tartaric acid (*32*) and also the just-mentioned pair H_A, H_C [or H_B, H_D] in cyclobutanone.

Symmetry elements of the second kind other than σ may generate enantiotopic ligands. Thus compound *42* in Fig. 16 (F and Ⅎ are enantiomorphic, i.e. mirror-image, ligands) has a center of symmetry only (C_i); H_A and H_B which are symmetry related by this center are enantiotopic. *meso*-Tartaric acid (*32*, Fig. 16) in what is probably its most stable conformation also has a center rather than a plane of symmetry, so that its enantiotopic methine hydrogens (H_A, H_B) are related by the i operation. Similarly, the four tertiary hydrogens in *43*, Fig. 16 are interrelated by the lone symmetry element S_4 in that molecule, and thus H_A is enantiotopic with H_B and H_D and the same is true of H_C. However, since the molecule also has a (simple) C_2 axis, H_A and H_C are homotopic, as are H_B and H_D. It might be noted here and in compound *34*, Fig. 13 that, while there can never be more than two enantiomers, a single ligand can have more than one enantiotopic partner.

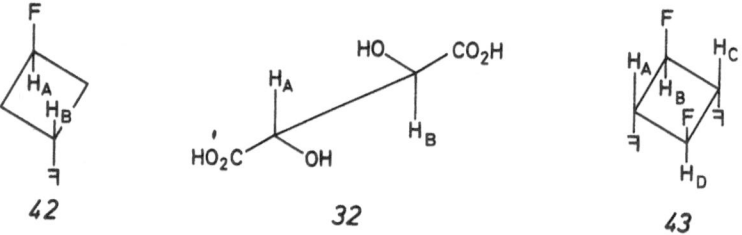

Fig. 16. Enantiotopic ligands in molecules with center or alternating axis of symmetry

Enantiotopic faces (Fig. 15) are also related by a symmetry plane — e.g. the plane of the double bond in *40*. The faces must not be interchangeable by operation of a symmetry axis, lest they be homotopic rather than enantiotopic.

The two faces in question may relate to a molecular plane other than the face of a double bond. Thus the two faces of methyl ethyl sulfide, $H_3C—S—C_2H_5$ are enantiotopic [5], inasmuch as oxidation of one or the other by peracid will give rise to two different, enantiomeric sulfoxides, (R)- and (S)-$CH_3SOC_2H_5$. Alternatively, we may consider this case as one of two enantiotopic ("phantom" [1]) ligands, namely the two unshared pairs on sulfur. Again the two faces of the benzene ring in 1,2,4-trimethylbenzene (*44*) are enantiotopic: addition of a chromium tricarbonyl ligand to one or other of the two faces gives enantiomeric coordination compounds *45a, b* (Fig. 17).

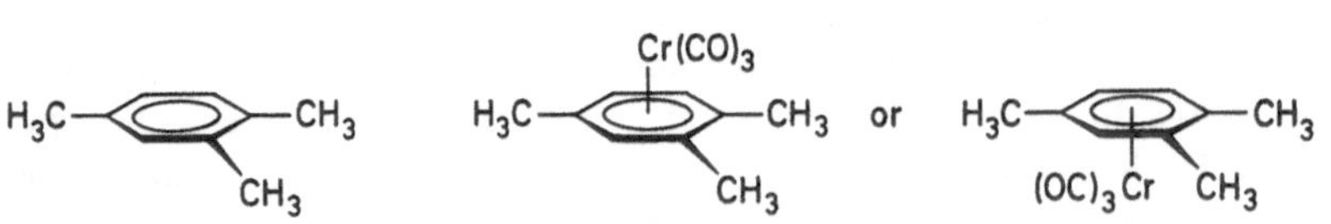

Fig. 17. Enantiotopic faces in substituted benzenes

So far we have discussed groups which are enantiotopic by internal comparison. Groups may also be enantiotopic by external comparison, i.e. groups in two different molecules are enantiotopic if they are related by reflection symmetry. Clearly this can be so only if the two molecules themselves are enantiomeric: corresponding groups in enantiomeric molecules (e.g. the methyl groups in D- and L-alanine) are enantiotopic. (See also Sect. 3.3.2, p. 19.)

Just as enantiomeric molecules cannot be distinguished in achiral environments, neither can enantiotopic ligands. Such ligands can, however, be distinguished by NMR spectroscopy in chiral media [26,27] or in presence of chiral shift reagents (discussed in Sect. 4.1), in synthetic transformations involving either chiral reagents or other types of chiral environment (asymmetric syntheses [28,29]) and, above all, in enzymatic reactions (since the enzyme catalysts are chiral) — cf. Section 5. It is because of these potential distinctions between enantiotopic ligands and faces that it is important to be able to recognize them.

3.3 Diastereotopic Ligands and Faces

The earlier mentioned criteria may also be employed to spot diastereotopic ligands, i.e. ligands which find themselves in a stereochemically distinct but non-mirror-image environment.

3.3.1 Substitution-Addition Criterion

In Fig. 18 are shown a number of cases where substitution of first one and then another of two homomorphic ligands by a different achiral test ligand (see footnote 5 on p. 9) gives rise to diastereomeric products.

Such ligands are called diastereotopic and are generally distinct both chemically and spectroscopically (their NMR signals will generally be different — cf. Section 4 — and their reactivity will, in general, be unequal).

The case of *46* (Fig. 18) is straightforward: since C(2) in 2-bromobutane is chiral, H_A and H_B cannot be enantiotopic and the replacement criterion discloses that they are diastereotopic rather than homotopic. The examples of cyclobutanol (*48*) and 4-*t*-butyl-1,1-difluorocyclohexane (*52*) show (cf. Sect. 1) that a chiral center is not required for the existence of diastereotopic nuclei. H_A and H_B in *48* and F_A and F_B in *52* are diastereotopic because they are *cis* and *trans*, respectively to the hydroxyl group at C(1) in *48* or the *t*-butyl group at C(4) in *52*. It might be noted that, after replacement, C(3) in *49* or C(1) in *53* is not a chiral center but a stereogenic center; the corresponding atoms in *48* and *52* are prostereogenic. In the case of propene (*50*) replacement of H_A and H_B generates a cis-trans pair of (diastereomeric) olefins again making H_A and H_B diastereotopic. (One is cis to the methyl group at the distal carbon atom, the other trans.) The case of *meso*-2,4-pentanediol (*54*) is of interest because the products of replacement of H_A and H_B are diastereomeric meso forms in which C(3) is pseudoasymmetric. C(3) in the progenitor molecule is called pro-pseudoasymmetric; in any case, H_A and H_B are diastereotopic. In diethyl sulfoxide, *56*, H_A and H_B are also diastereotopic as are H_C and H_D; on the other hand, H_A is enantiotopic with H_C and H_B with H_D. (Since the molecule as a whole is achiral, the existence of enantiotopic atoms is possible.) The situation in *56* is entirely analogous to that in citric acid (*14*, Fig. 4) in which we now recognize H_A to be diastereotopic to H_B (as H_C is to H_D) whereas H_A and H_C (or H_B and H_D) are enantiotopic. Citric acid and diethyl sulfoxide are said to contain two enantiotopic pairs of diastereotopic

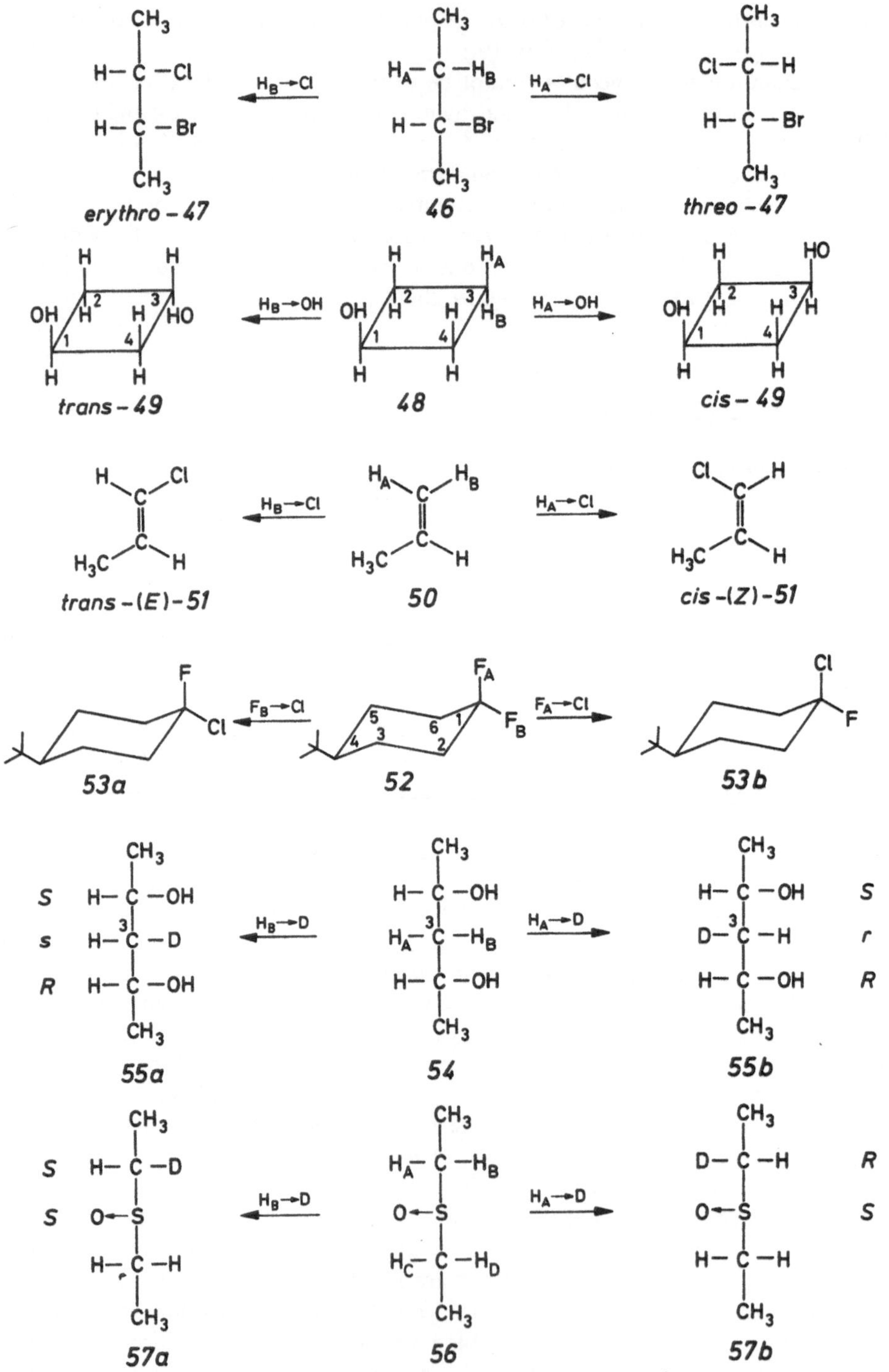
erythro-47
46
threo-47
trans-49
48
cis-49
trans-(E)-51
50
cis-(Z)-51
53a
52
53b
55a
54
55b
57a
56
57b

Fig. 18. Diastereotopic ligands

ligands (or two diastereotopic pairs of enantiotopic ligands). In passing we may note that the hydrogens attached at C(2) and C(4) in cyclobutanol (*48*, Fig. 18) also form enantiotopic pairs of diastereotopic hydrogens. In the trans diol (*trans-49*), on the other hand, the corresponding hydrogens form an enantiotopic pair [C(2) vs. C(4)] of geminally homotopic hydrogens whereas in the cis 1,3-diol (*cis-49*) they form a

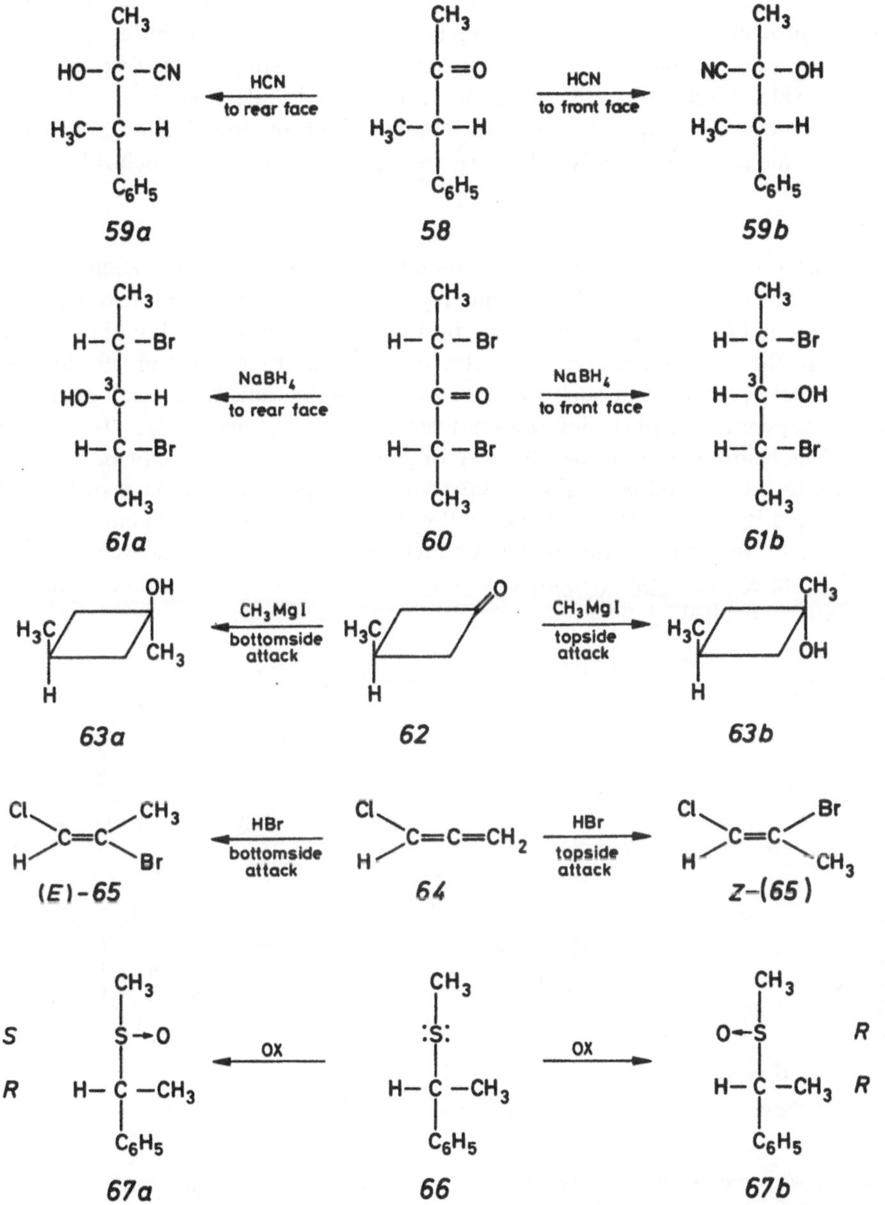

Fig. 19. Diastereotopic faces of double bonds

geminally diastereotopic pair of homotopic [C(2) vs. C(4)] hydrogens. (These facts may become clearer when symmetry criteria are applied, see below.)

The addition criterion may similarly be applied to recognize diastereotopic faces. Methyl α-phenethyl ketone, *58* in Fig. 19 has a chiral center; addition clearly gives rise to diastereomers (*59a, 59b*); the faces of the carbonyl carbon are diastereotopic and the C=O group is prochiral. This case is of importance in conjunction with Cram's rule [10]. Compounds *60, 62* and *64* also display diastereotopic faces even though the products *61, 63* and *65* are not chiral; *60, 62* and *64* have prostereogenic rather than prochiral faces. The C=O group in *60* is propseudoasymmetric, since C(3) in *61* is a pseudoasymmetric center. α-Phenethyl methyl sulfide (*66*) displays diastereotopic sides of a molecular plane not due to a double bond [5, 24] and may alternatively be considered a case of diastereotopic phantom ligands (unshared pairs on sulfur). This case does involve chirality and the sulfur atom is prochiral.

3.3.2 Symmetry Criterion

The symmetry criteria of diastereotopic ligands or faces are simple: such ligands or faces must be related neither by a symmetry element of the first kind (axis) nor by one of the second kind (plane, center, alternating axis).[8] The reader should convince himself that the even-numbered molecules depicted in Figs. 18 and 19 (middle column) are either devoid of such symmetry elements or that, when such elements (e.g. σ) are present, their operation does not serve to interchange the ligands or faces designated as being diastereotopic. By way of generalization [24] it might be pointed out that *54* in Fig. 18 and *60* in Fig. 19 correspond to the general type A in Fig. 20 (diastereotopic ligands — R — or faces). Type B has homotopic ligands or faces as does type C. In contrast, the ligands or faces in D and E are diastereotopic [24]. Those in F are enantiotopic. (F and Ⅎ stand for enantiomorphic, i.e. mirror-image ligands, X is an achiral ligand.)

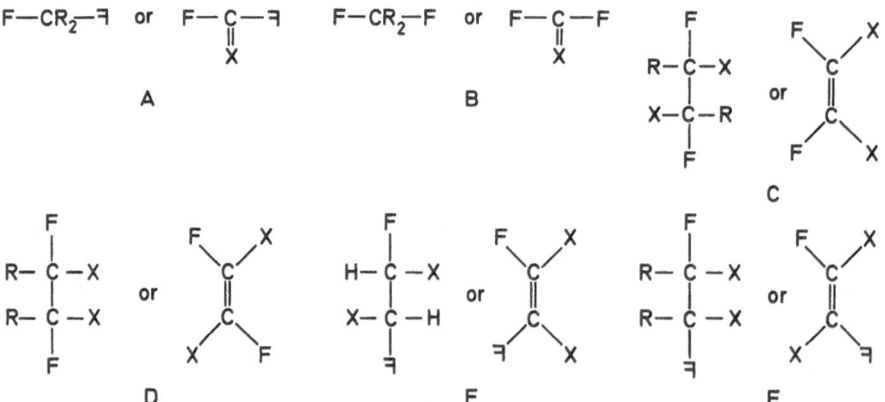

Fig. 20. Topicity of ligands and faces in molecules with paired chiral substituents

---

[8] It is also to be understood that the ligands or faces in question must be in constitutionally equivalent environments (such as C-2 and C-4 in *n*-pentane). If this is not the case (e.g. as between C-2 and C-3 in *n*-pentane), one speaks of constitutional heterotopicity; cf. Fig. 21 and p. 19.

Ligands may be diastereotopic by external as well as internal comparison. Corresponding ligands in diastereomers are diastereotopic under any circumstances; corresponding ligands in enantiomers are diastereotopic when viewed in a chiral environment (e.g. a chiral solvent) [26,27]. It has been pointed out, however [30], that the difference between internally and externally heterotopic nuclei is of no fundamental significance (though it is probably worth retaining as a matter of convenience). For example, when one compares ^{13}C NMR signals of enantiotopic or diastereotopic methyl groups, e.g. in an appropriately placed isopropyl moiety, Me_2CH, one generally creates the fiction that the two ^{13}C labeled methyl groups are in the same molecule, i.e. internally stereoheterotopic. In fact, however, if the spectrum is recorded with material that is not isotopically enriched, i.e. in which ^{13}C occurs at the natural abundance of 1.1%, the chance that two adjacent methyl groups are both ^{13}C labeled is only about 1 in 10,000 and such molecules are not ordinarily seen in the spectrum. The *actual* comparison is therefore between the enantiomeric or diastereomeric molecules, e.g. $^{13}Me_aCHMe_bX$ and $Me_aCH^{13}Me_bX$ where X is an achiral or a chiral group, as the case may be, and the superscript 13 marks the methyl group which contains the isotope; thus it is actually a comparison between *externally* enantiotopic or diastereotopic groups [30]!

3.4 Concepts and Nomenclature [4,5,9]

At this point it may be well to review the criteria for homotopic, enantiotopic and diastereotopic ligands or faces. Ligands, or faces, are equivalent or homotopic when they can be brought into coincidence by operation of a proper (C_n) symmetry axis. If this condition is not fulfilled but the ligands or faces can be brought into coincidence by operation of an improper (S_n) axis of symmetry, including a plane (σ) or center (i) of symmetry, the ligands or faces are enantiotopic. If neither symmetry operation (C_n or S_n) brings the ligands or faces into coincidence, they are diastereotopic or (see footnote 8 on p. 18) constitutionally heterotopic. In the realm of external comparison, molecules have homotopic groups if they are superposable, enantiotopic groups if they are enantiomeric. Corresponding groups in diastereomeric molecules are diastereotopic.

It is illuminating to make a comparison between isomeric and nonisomeric compounds on the one hand and homotopic or heterotopic ligands or faces on the other. There is logic to such a procedure since it was explained earlier that stereoisomers are generated by appropriate replacement of heterotopic ligands or addition to heterotopic faces. Figure 21 displays such a comparison [6]. It is convenient, in conjunction with the diagram of homotopic and heterotopic ligands (Fig. 20) to introduce an additional term: If homomorphic ligands (e.g. the hydrogen atoms in a methylene group) occur in constitutionally distinct portions of a molecule, we call them constitutionally heterotopic. Examples would be the methylene hydrogens at C(2) and those at C(3) in cyclobutanol (*48*, Fig. 18).

Constitutionally heterotopic ligands are in principle always distinguishable, just as constitutional isomers are. Diastereotopic and enantiotopic ligands or faces may be lumped together under the term "stereoheterotopic" just as diastereomers and enantiomers are both called stereoisomers.

Ernest L. Eliel

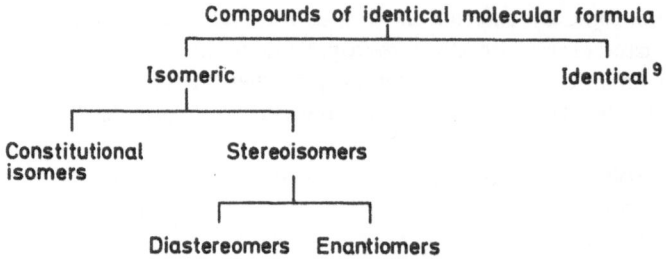

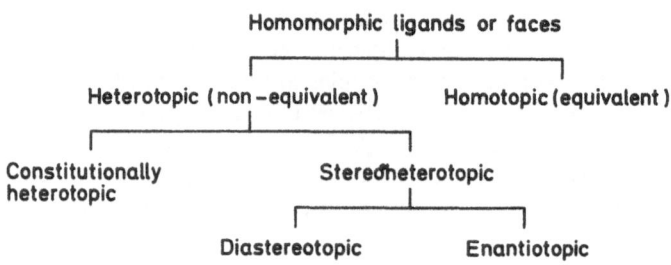

Fig. 21. Classification of compounds and ligands [6]

Just as it is convenient to distinguish enantiomers and diastereomers by nomenclature symbols (*R, S, E, Z*, etc.) it is desirable to provide names for stereoheterotopic ligands or faces. The basic nomenclature to this end has been provided by Hanson [4,6] and is closely related to the nomenclature of stereoisomers.

If, in a prochiral assembly — e.g. a prochiral center Caabc — a hypothetical precedence (in the sense of the sequence rule [1]) is given to one of the identical ligands (a) over the other (a'), that ligand (a) will be called "*pro-R*" if the newly created "chiral center" Caa'bc (sequence a > a') has the *R* configuration, but it will be called "*pro-S*" if the newly created "chiral center" has the *S* configuration. Let us take ethanol (*68*, Fig. 22) as an example. The hydrogen atoms H_A and H_B are enantiotopic (Sect. 3.2). If preference is given to H_A over H_B in the sequence rule, the sequence is OH, CH_3, H_A, H_B and the (hypothetical) configurational symbol for *68* would be *R*, hence H_A is *pro-R*; by default, H_B is *pro-S*. The answer would have come out the same if H_B has been given precendence over H_A; in that case the

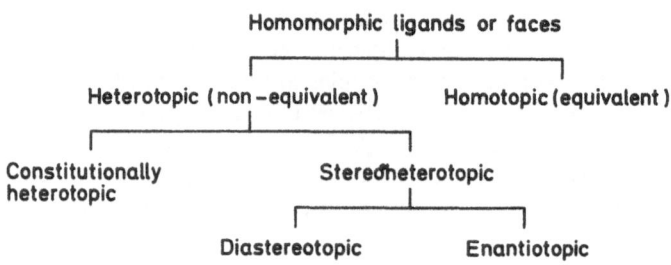

Fig. 22. Ethanol, (*R*)-ethanol-1-*d* and (*S*)-ethanol-1-*d*

---

[9] The term "homomeric" has been proposed for such species by Mislow, K.: Bull. Soc. Chim. Belg., *86*, 595 (1977) and O'Loane, J. K.: Chem. Revs., *80*, 41 (1980).

20

sequence would have been OH, CH_3, H_B, H_A and the hypothetical configurational symbol for *68* is then S — hence H_B is *pro-S*. It might be noted (cf. Fig. 22) that the same result would obviously have been obtained by replacing first one hydrogen and then the other by deuterium since deuterium has sequential precedence over hydrogen[1]; replacement of H_A by D gives (*R*)-ethanol-1-*d* [(*R*)-*69*] and hence H_A is *pro-R*; similarly, replacement of H_B by deuterium gives (*S*)-ethanol-1-*d* [(*S*)-*69*] and hence H_B is *pro-S*. This alternative of replacing the atom to be stereochemically labeled by a heavy isotope rather than giving it hypothetical precedence may be used except when the heavy isotope is already present at the prostereogenic center. Thus the enantiotopic methyl hydrogens of acetic-2-*d* acid, CH_2DCO_2H can obviously not be assigned as *pro-R* or *pro-S* by replacing them by deuterium (see footnote 5 on p. 9). In this case it is necessary to elevate one or other of the hydrogens in question to a precedence above its counterpart but below D.

In a formula, the "*pro-R* group X" is sometimes written as X_R (and similarly X_S for the *pro-S* group). It is important, however, to read X_R as "the *pro-R* group X" and not as the "*R* group X" since prochirality not chirality is implied. Indeed, Fig. 23 shows a case (type CFFXY) where both CH_3CHOH (F) ligands have the *S*-configuration but the upper one is *pro-R* whereas the lower one is *pro-S*. In the original structure, the central atom [C(3)] is achiral. When precedence is given to the upper ligand (indicated in Fig. 23 by replacement of CH_3 by $^{13}CH_3$), however, C(3) becomes chiral and since its configuration is *R*, the upper CH_3CHOH ligand is *pro-R*[31].

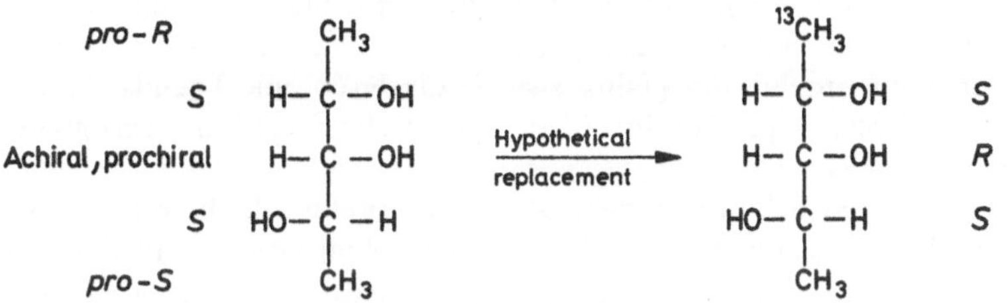

Fig. 23. Molecule in which ligand of *S*-configuration is *pro-R*

Just as chiral centers can be labeled *R* or *S* not only in enantiomers but also in many diastereomers, so the designations *pro-R* and *pro-S* are not confined to enantiotopic ligands but may also be used for a number of diastereotopic ones (for exceptions, see below). Thus, for example, the labeling in Fig. 13 is such that H_A (compounds *30, 32, 34, 36*) or Me^1 (compound *38*) is the *pro-R* group; the reader should verify this proposition. The same is true for compounds *46* and *56* in Fig. 18. Compounds *48, 50, 52* and *54* in Fig. 18 cannot be labeled in this manner since replacement of the diastereotopic ligands does not produce chiral products. In *54* (pro-pseudoasymmetric center) the substitution gives rise to a pseudoasymmetric center which, in the compound of the left is *s*, in the compound on the right *r*. Hence H_A is called *pro-r* and H_B *pro-s*[6].

In *50*, replacement of H_A and H_B gives rise to Z and E olefins, respectively;[10] here H_A should be called [6] "*pro-Z*" and H_B "*pro*-E". (Symbols H_r, H_s, H_E, H_Z may be used.) In compounds *48* and *52*, the terms "*pro-cis*" for H_A and F_a and "*pro-trans*" for H_B and F_B (symbols H_{cis}, H_{trans}, etc.) are appropriate.

Hanson [4] has also devised a specification of heterotopic faces. The rule here is simple: one looks at the chirality in two dimensions (cf. Sect. 6) and if the sequence is clockwise, one calls it *Re*, if counter-clockwise, *Si*.[11] Thus the face of acetaldehyde turned toward the reader in Fig. 15 is *Si* (O, CH_3, H are in counter-clockwise order) and the corresponding front faces in Fig. 19 are *Re* for *58* and *re* for *60* (here the face is pro-pseudoasymmetric and the use of the lower-case symbol is appropriate). The nomenclature does not work for *62* and *64*[12] but it would be appropriate here to call the top face "*ci*" and the bottom face "*tr*", these being the first two letters of cis and trans, respectively. (*Re* and *Si* are, of course, the first two letters of *Rectus* and *Sinister*.) In *66* (Fig. 19) the uninvolved lone pair must be inserted as a phantom ligand; when this is done the right face of the molecule becomes *Si*, the left one *Re*. (As already mentioned, one may consider both lone pairs as prochiral ligands, in which case the right pair is *pro-S*, the left *pro-R*, since elevation of the right pair over the left gives a hypothetical S configuration, and vice versa. (The fact that attachment of oxygen to the right pair gives an S-sulf-oxide — and to the left pair the R isomer — is immaterial in either system of nomenclature.)

We have already mentioned (p. 4 and Fig. 3) that prostereoisomerism can also exist in cases where replacement of one of two homomorphic ligands gives rise to molecules of axial or planar chirality. Compounds *10* in Fig. 3 and *36* in Fig. 13 are examples of axial prochirality giving rise to enantiotopic ligands; compound *12* in Fig. 3 is an example of planar prochirality giving rise to such ligands. Figure 24 shows examples of axial prochirality giving rise to diastereotopic ligands [33], viz. *70* [34] and *71* [35] and of planar prochirality (if it may be considered as such) giving rise to such ligands, as in *72* [36].

Although systematic nomenclature is generally to be preferred, there are some instances, for example in steroids, where a local or parochial nomenclature is still generally used. Thus in 3-cholestanone (*73*, Fig. 25) the hydrogen atoms above the plane of the paper (which itself represents a projection of the three-dimensional

---

[10] For nomenclature purposes, the replacement of H by Cl is not appropriate; one should replace H by D or better (see above) by "elevated H". In the case of *50*, Fig. 18, it happens to make no difference.

[11] Hanson [4] originally used *re* and *si*, but since prochiralty, not propseudoasymmetry, is involved, the use of capital letters is more appropriate and has now been accepted [11,32].

[12] (Added in proof) With the recent additions to stereochemical nomenclature [1b], it becomes possible to name the faces of compound *62* in *re/si* terminology. The cyclobutanone is opened with a double complementation of C(1) yielding (i) in which the left branch has auxiliary descriptor R_o, the right S_o. The top face is thus *si* and the bottom face *re* [178].

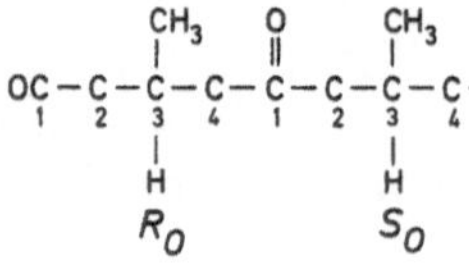

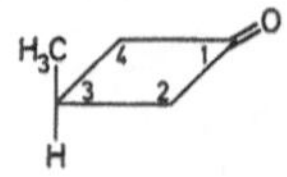

CH(CH₃)₂ → $CH(\underline{C}H_3)_2$

$C_2H_5CHBr-CO-C(Me)=C=C\underline{H}_2$

70

$(\underline{C}H_3)_2\,CHCR=C=CR'\,R''$

71

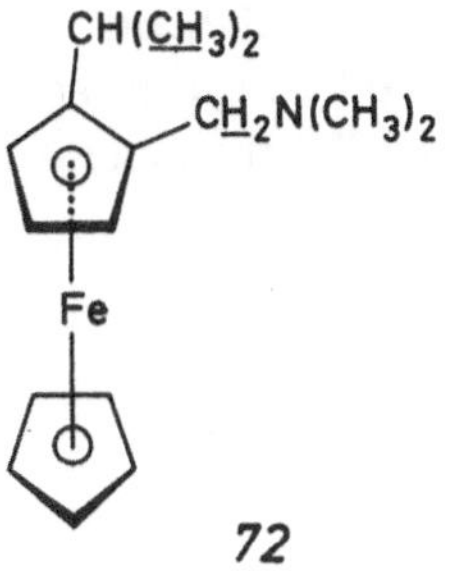

$CH(\underline{C}H_3)_2$
$CH_2N(CH_3)_2$
Fe

72

Fig. 24. Prochiral axes and planes

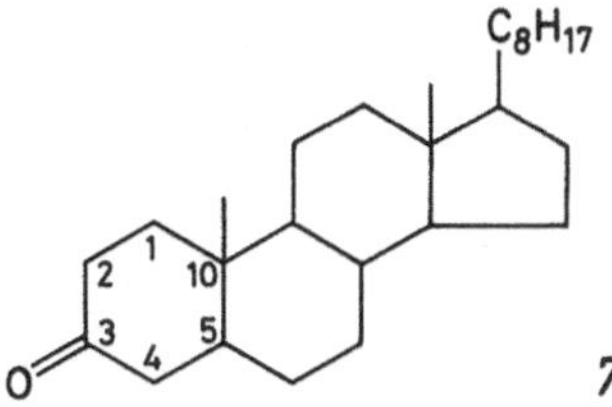

C_8H_{17}

73   **Fig. 25.** 3-Cholestanone

molecule) are called β and those below the plane α [37]. Since the geminal hydrogens at each methylene carbon form a diastereotopic pair, it is clear that diastereotopic hydrogen atoms in such pairs may be distinguished by calling them H_α and H_β and this is commonly done. There is obviously no one-to-one connection of such common with systematic nomenclature; for example, the β-hydrogen at C(2) is *pro-S* but that at C(4) is *pro-R*. Not surprisingly this lack of correlation parallels that between α/β and R/S when one looks at chiral centers in steroids (as in 1- and 4-cholestanol). The α- and β-designation may also be given to prostereoisomeric faces; thus the front (*Si*) face of the keto function at C(3) is β and the rear (*Re*) face is α.

Sometimes, especially in enzymology, it is convenient to speak of the face of a molecule quite apart from any particular prochiral or prostereogenic element. For example, one might like to express the fact that a steroid is attached to an enzyme receptor on the α face without making reference to any particular CH_2 or C=O moiety. The *pro-R/pro-S* or *Re/Si* nomenclature is not generally applicable to such cases (for an exception, see Fig. 17). The α/β face nomenclature (which applies to steroids as explained above) has been generalized [38] to apply to all kinds of rings using the following rules:

1) If the compound is monocyclic, or if the rings are not fused, the faces are designed as α if progression around each ring from the lowest to the next higher numbered atom by the shortest route is *clockwise*, as β if such progression is *counterclockwise*. The ring is to be examined as a planar regular polygon, i.e. disregarding conformation. If multiple numbering systems are used, the precedence is 1→2 > 1a→2a > 1′.

2) If the compound is made up of "ortho-fused" rings only, face designations of the entire system of rings are derived from the ring containing the lowest numbered unshared atoms, as specified in standard numbering for the compound.

Under rule 2, the A ring is pace-setting for the entire steroid system (Fig. 25) and under rule 1 the front face of this ring is β since the numbering proceeds counterclockwise. Similarly the front face of heme (Fig. 26) is α; oxygen in hemoglobin and myoglobin is known to bind to the β-face [39]. For further details the original publication should be consulted [38]; the system has not yet been generally accepted.

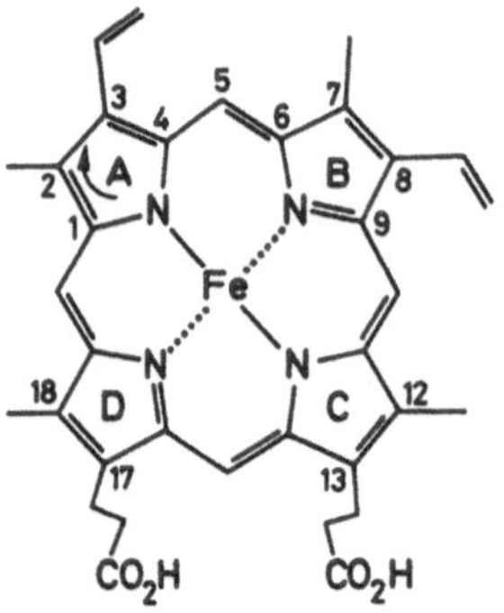

Fig. 26. Faces of heme

4 Prochirality and Nuclear Magnetic Resonance [5,40,41]

4.1 General Principles. Anisochrony

Nuclei which are diastereotopic will, in principle, differ in chemical shift, i.e. they will be "anisochronous" [5].[13] It must be pointed out, however, that the chemical shift differences are often small, sometimes so small that the signals can be resolved only at quite high fields, and not infrequently altogether unobservable even at the highest fields available. In the latter situation one may speak of "accidental isochrony", meaning that while the nuclei are in principle anisochronous, they are not, in fact, resolved.

Anisochrony for diastereotopic ligands is seen with a number of different nuclei. We have already mentioned that $CH_2BrC(CH_3)BrCO_2CH_3$ displays different signals for the diastereotopic protons (italicized) [21] and that $CF_2BrCHBrC_6H_5$ displays different resonances for the diastereotopic fluorine nuclei [20]. The diastereotopic methyl groups in the ferrocenyl cation 74 (Fig. 27) are distinct both in their ^{1}H and ^{13}C signals [42].

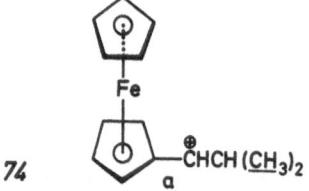

74

Fig. 27. Example of diastereotopic methyl groups

---

[13] This term was coined by G. Binsch following A. Abragam's use of "isochronous" for chemical-shift equivalent; cf. Ref. 5, p. 23.

The immediate cause for anisochrony is, of course, the unequal magnetic field sensed by the diastereotopic nuclei. It follows that as the source of the diastereotopic environment is removed further and further from the test nuclei, the anisochrony is expected to diminish. This prediction has been tested [43] with the results shown in Table 1.

Table 1.[43] Observed Anisochrony of Diastereotopic Methyl Protons in $(CH_3)_2CH-X-CH(CH_3)C_6H_5$

Entry	X	Shift Difference	
		in ppm	
		in CCl_4	in C_6H_6
1	none	0.182	0.133
2	O	0.067	0.013
3	OCH_2	0.005	0.008
4	OCH_2CH_2	0.042	0.030
5	OCH_2CH_2O	0.000	0.013
6	$OCH_2CH_2OCH_2$	0.000	0.000

It is seen that, in general, the above proposition is correct but an exception is found for the OCH_2CH_2 "spacer" (entry 4) which leads to greater anisochrony than does OCH_2. This may be an effect of coiling (contribution of gauche conformations, enhanced by the presence of oxygen in the chain) which brings the chiral end, $-CH(CH_3)C_6H_5$ close to the diastereotopic methyl groups, thus enhancing the magnetic field difference of these groups over what it is in the lower homolog (CH_2O spacer, entry 3). It is also seen in Table 1 that anisochrony in benzene may be either greater or smaller than in carbon tetrachloride; in particular, in entry 5 anisochrony is observable in benzene but not in CCl_4. It is thus desirable, in looking for potential anisochronies, to record the spectrum in several different solvents [44] (CCl_4, $CDCl_3$, benzene-d_6, pyridine-d_5) (see also [40, 45]). Another way of enhancing (or manifesting) anisochronies is to use lanthanide shift reagents [44].

In Table 2 are summarized the shift differences (both ^1H and ^{13}C) between the diastereotopic methyl groups of the compounds [46] shown in Fig. 28. (Arguments are adduced in the paper [46] that the conformation shown is by far the preferred one, at least for R = COX.) It is immediately obvious that these differences in shift between diastereotopic protons are much larger for the phenyl than for the cyclohexyl compound; presumably because of the much larger differential shielding of the

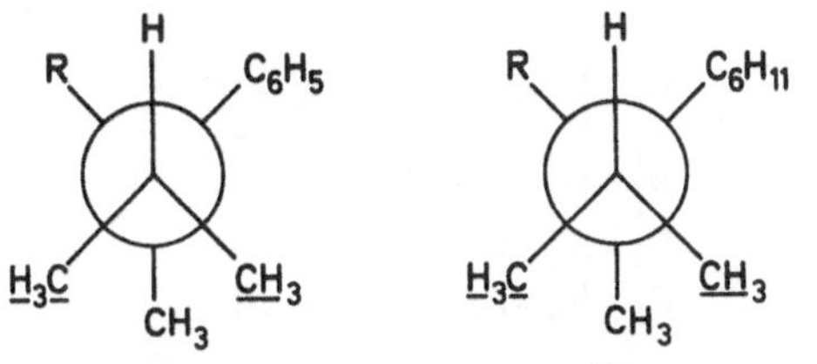

Fig. 28. Preferred conformations of $(CH_3)_2CHCR(CH_3)C_6H_5$ (75) and $(CH_3)_2CHCR(CH_3)C_6H_{11}$ (76)

methyl protons by the phenyl ring. On the other hand, these ^{1}H shift differences (except in the case of the nitrile) are not strongly dependent on the nature of R. (Neither of these statements is true for the ^{13}C shifts of the phenyl compounds which, in some instances, are opposite to the ^{1}H shifts.) It is clear, also, that the anisochronies of the diastereotopic methyl groups are much larger in ^{13}C than in ^{1}H resonance; presumably this is a consequence of the generally larger shift effects in ^{13}C NMR spectra due to their largely paramagnetic origin. ^{13}C NMR is thus generally a better probe for diastereotopic ligands than ^{1}H where both types of nuclei are present in these ligands.

Table 2. Chemical Shift Differences (ppm) of Diastereotopic Groups in Compounds *75* and *76*

R	CO_2H	CO_2CH_3	CO_2Ph	COCl	$COCH_3$	$CONH_2$	CH_2NH_2	CH_2OH	CH_2OAc	CN
75, ^{1}H	0.39	0.35	0.41	0.38	0.31	0.27	0.29	0.27	0.20	−0.37
76, ^{1}H	0.07	0.06	0.00	0.06	0.00	0.06	0.03	0.04	0.00	0.11
75, ^{13}C	1.84	1.90	1.79	0.91	1.14	0.94	−0.80	−0.54	−0.14	0.71
76, ^{13}C	0.85	0.85	0.80	0.50	0.74	0.84	0.42	0.51	0.33	1.67

The diastereotopic and hence anisochronous nuclei so far considered were attached to prochiral centers. Mention was made earlier of axial and planar prochirality which may also give rise to diastereotopic nuclei. A case of planar prochirality [47] leading to anisochronous nuclei is shown in Fig. 29. Simultaneous complexation of (*R*)-methyl *p*-tolyl sulfoxide and propene to PtCl$_2$ gives rise to diastereomers (*77a, b*, depending on which enantiotopic face of the propene is turned toward the metal) in which the propene methyl groups are (externally) diastereotopic and hence anisochronous. Similar complexation with *cis*-2-butene (in which the two faces are homotopic) gives a single enantiomer (*78*), but now the two methyl groups of the butene are internally diastereotopic and once again anisochronous. A surprising result is seen with *trans*-2-butene: here the faces are again enantiotopic and so two diastereomers (*79a, b*) should be formed, but within each diastereomer one would expect the two methyl groups (of one and the same complex) to be homotopic, because of their interchangeability by the C$_2$ axis. Thus one would expect to see two (externally) diastereotopic methyl signals, each corresponding to two methyl groups. In fact, however, *four* methyl signals are seen, i.e. there is no C$_2$ axis [47]. This has been

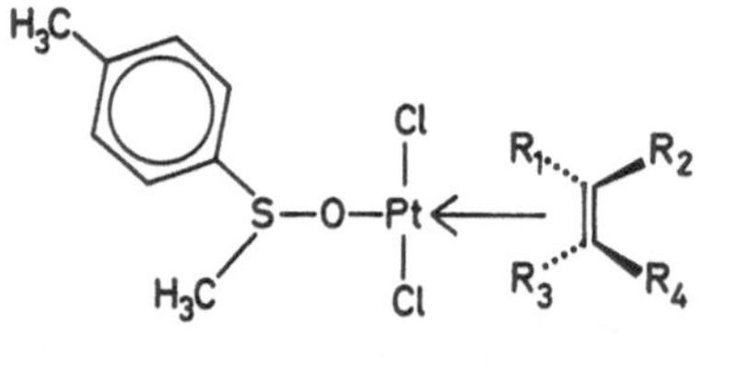

77a $R_1 = CH_3, R_2 = R_3 = R_4 = H$
77b $R_1 = R_3 = R_4 = H, R_2 = CH_3$
78 $R_1 = R_3 = H, R_2 = R_4 = CH_3$
79a $R_1 = R_4 = CH_3, R_2 = R_3 = H$
79b $R_1 = R_4 = H, R_2 = R_3 = CH_3$

Fig. 29. Anisochrony due to planar prochirality

ascribed to slow rotation, on the NMR time scale, around the olefin-platinum bond [47].

The complexation of dimenthyl maleate and dimenthyl fumarate with iron tetra-carbonyl (Fig. 30) [48] gives rise to two diastereomers in the case of the fumarate (whose olefin faces are diastereotopic) but only a single enantiomer in the case of the maleate (whose faces are homotopic). Nevertheless, both complexes display diastereotopic olefinic protons: the maleate complex because the two ethylenic protons are internally diastereotopic, the fumarate complex because, although the two protons within one molecule are homotopic by virtue of the existence of a C_2 axis, the olefinic protons of the diastereomeric molecules are externally diastereotopic. One may ask, then, whether the two cases are distinguishable, and indeed they are. The two protons in the maleate complex (80) will necessarily be equally intense but, being anisochronous, will split each other and thus give rise to an AB signal. The two diastereomers in the fumarate complex (81a, b) are not necessarily formed in equal amounts and therefore their signals may be unequal in intensity. Moreover, since the protons within one molecule are homotopic, they do not split each other and one thus sees two (possibly unequally intense) singlets.

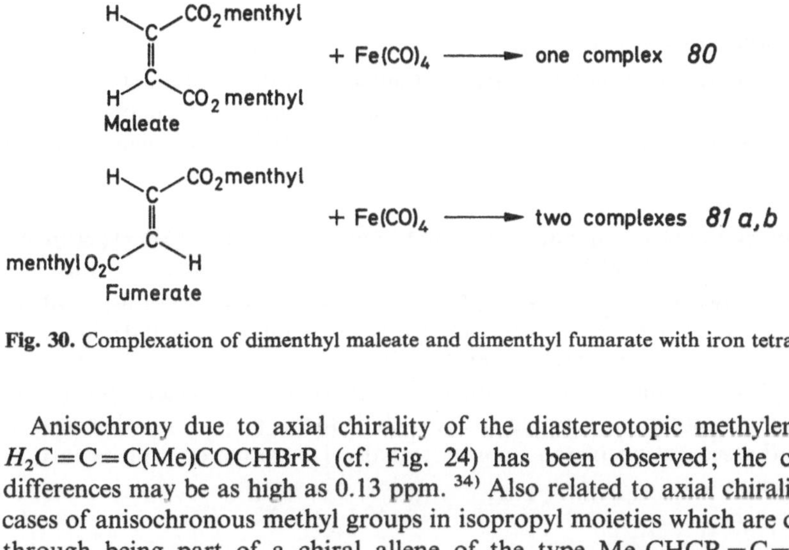

Fig. 30. Complexation of dimenthyl maleate and dimenthyl fumarate with iron tetracarbonyl

Anisochrony due to axial chirality of the diastereotopic methylene protons in $H_2C=C=C(Me)COCHBrR$ (cf. Fig. 24) has been observed; the chemical shift differences may be as high as 0.13 ppm. [34] Also related to axial chirality are several cases of anisochronous methyl groups in isopropyl moieties which are diastereotopic through being part of a chiral allene of the type $Me_2CHCR=C=CR'R''$ [35,49]. These cases resemble that shown in Fig. 27 where a prochiral center ($Me_2CH—C ...$) is attached to a chiral ferrocenyl moiety; it should be noted that the ferrocenylmethyl-carbenium ion fragment is chiral only if rotation about the $C_p—C^+$ bond (marked a in Fig. 27) is slow on the NMR time scale [42].

Since nuclear magnetic resonance is a scalar probe, enantiotopic nuclei are isochronous (i.e. have the same chemical shift) in achiral media. Such nuclei, however, become diastereotopic in chiral media and thus, in principle (though often not in practice) anisochronous. Among many examples [26,27] are the enantiotopic methyl protons of dimethyl sulfoxide, CH_3SOCH_3, which are shifted with respect to each other by 0.02 ppm [26c] in solvent $C_6H_5CHOHCF_3$. (Surprisingly the ^{13}C signals of the

27

two methyl groups are not resolved under these conditions; this is an exception to the rule that ^{13}C signals of diastereotopic methyl groups generally show larger relative shifts than their ^1H signals [46,50].

The methyl groups of dimethyl sulfoxide are also anisochronous in the presence of chiral lanthanide shift reagents, such as Eu(facam) or Eu(hfbc)$_3$ (Fig. 31) [51]. The enantiotopic carbinol protons of alcohols RCH$_2$OH are similarly rendered anisochronous by chiral shift reagents [52].

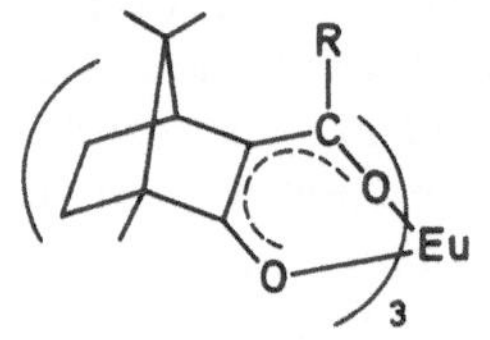

R = CF$_2$CF$_2$CF$_3$: Eu(hfc)$_3$ or Eu(hfbc)$_3$ [53d,e]

R = CF$_3$: Eu(tfc)$_3$ or Eu(facam)$_3$ or Eu-Opt ® [53c,e]

R = (CH$_3$)$_3$C [53a,b,e]

Fig. 31. Chiral shift reagents

NMR shift differences between groups which are enantiotopic by external comparison (i.e. in enantiomers) may likewise be induced by either chiral solvents [26,27] or chiral shift reagents [52]. Integration of the areas of signals of enantiomers so shifted is used for the determination of enantiomeric excess, a topic which cannot be taken up here but has been discussed elsewhere [53].

The detection of diastereotopic nuclei by NMR is possible only if the diastereotopic nature of such nuclei is maintained on the time scale of the NMR experiment. Thus the equatorial and axial fluorine atoms in 1,1-difluorocyclohexane (Fig. 32), though diastereotopic, give rise to a single NMR signal because the rate of interchange of these nuclei by ring reversal at room temperature (cf. 100,000 sec^{-1}) is much higher than the shift between the fluorine nuclei (884 Hz at 56.4 MHz or 884 sec^{-1}) [54]. However, the fluorine atoms F^1 and F^2 become anisochronous below —46 °C when interconversion between the two chair forms (Fig. 32) is slowed to a rate less than the separation of the fluorine signals. This situation will be further discussed in Section 4.4.

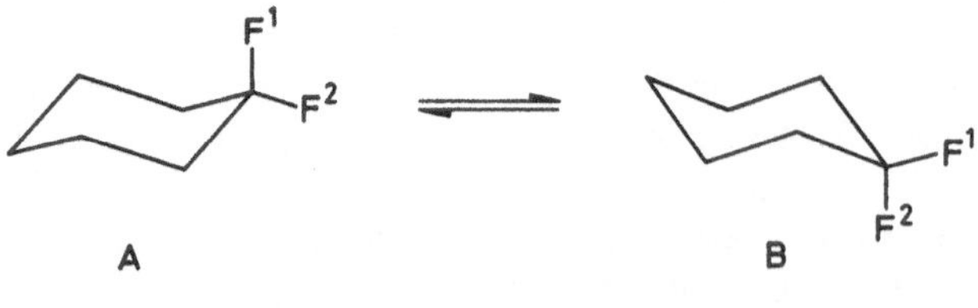

A   B   Fig. 32. 1,1-Difluorocyclohexane

In conclusion of this Section we want to mention the phenomenon of "multiple nonequivalence" [40,45] which may occur when there are several nuclei in a molecule which are diastereotopic to each other. Two cases are depicted in Fig. 33. In the allenic molecule *82* [49a)], the two ethoxy groups are diastereotopic because of the allenic chirality, and within each ethoxy group the methylene protons are similarly diastereotopic since they are not related by either a C_2 or a σ; as a result, all four methylene protons are anisochronous and there will be two sets of AB signals. The same is true of the four methylene protons in the biphenylic sulfoxide *83* [55)]. The symmetry plane which might normally make H_1 enantiotopic with H_2 (and H_3 with H_4) is absent because of the non-coplanarity of the biphenyl ring system. Multiple nonequivalence is also seen in malonates: in the methylene proton signals of $R^*CH(CO_2C\mathit{HH}R')_2$ [56)] and in the isopropyl methyl signals of $R^*CH(CO_2CH\mathit{Me}_2)_2$ (R^*) = chiral substituent).[57)]

82 *83*

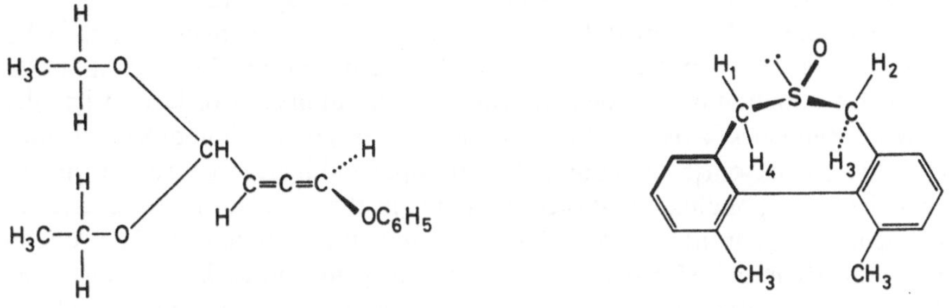

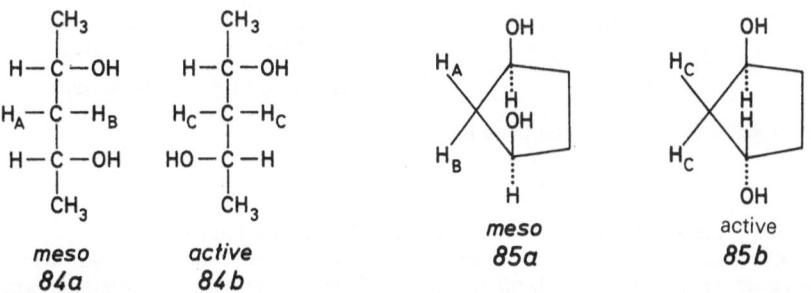

Fig. 33. Multiple nonequivalence

4.2 Configurational Assignment by NMR. Assignment of Prochirality Descriptors

In this Section we shall use the ideas of prochirality in assignment of stereochemical configuration [58)] (usually relative — especially meso vs. *dl* — rather than absolute configuration) and we shall also discuss assignment of prochirality symbol (i.e. recognition of which group is *pro-R* and which *pro-S* at a prochiral center). (Recognition of prochiral faces as *Re* or *Si* is usually obvious from the stereochemistry of the addition products thereto and will not be discussed here; examples are found in Section 5.2).

meso *84a* *active* *84b* *meso* *85a* active *85b*

Fig. 34. Distinction of active and meso forms by NMR

In favorable circumstances, active (or *dl*) and meso stereoisomers may be distinguished directly; an acyclic (*84*) and a cyclic (*85*) example are shown in Fig. 34. In both cases the methylene protons H_C in the active forms are related by a C_2 axis and therefore homotopic and isochronous whereas the corresponding protons H_A and H_B in the meso forms are not related by either C_2 or σ and are therefore diastereotopic and anisochronous. The situation is not altered when the *dl* form rather than an active isomer is compared with the meso form: the (internally homotopic) methylene protons of the two enantiomers are externally enantiotopic and so remain isochronous.

Cases where the stereogenic centers are further removed from the prostereogenic one — e.g. in compounds of type Ph—CHX—CH$_2$—CMe$_2$—CH$_2$—CHX—Ph — have also been investigated in both acyclic [59] and cyclic [60] systems.

When no suitable probe is present, as in Fig. 35, such a probe may sometimes be introduced. There are two ways of doing this, depicted in Figs. 35 and 36, respectively. Introduction of two identical chiral groups at ligands or faces originally related by a symmetry axis or plane leaves the originally homotopic ligands CH$_3$CH in the active isomer *86a* (Fig. 35) homotopic and hence isochronous in *87a* but makes the corresponding enantiotopic ligands in the meso form *86b* diastereotopic and hence anisochronous in *87b*. Unfortuantely, the usefulness of the probe in this form is greatly impaired because the homotopic groups in each enantiomer (*R*- and *S*-*86a*) become diastereotopic[14] (by external comparison) in the pair *87a*. Therefore one cannot distinguish the *dl* pair (*RS-86a*) from the meso form (*86b*); distinction of active from meso forms is, of course, effected more simply by polarimetry.[15]

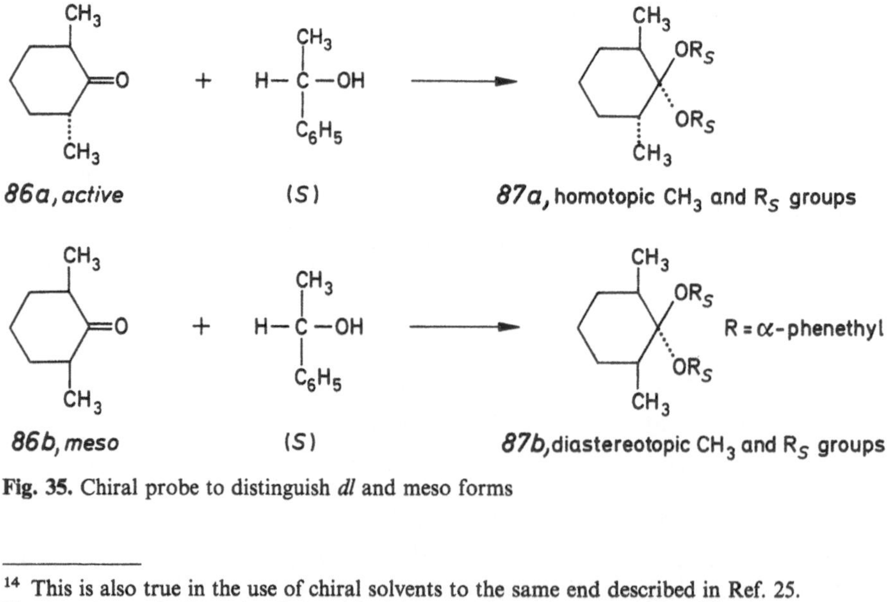

Fig. 35. Chiral probe to distinguish *dl* and meso forms

---

[14] This is also true in the use of chiral solvents to the same end described in Ref. 25.

[15] *dl-86a* can, of course, be distinguished from meso-*86b* by the classical [10] method of introducing a new chiral center, e.g. by reducing the ketone to an alcohol: *86b* will give rise to two diastereomers, *86a* to only one. See also Ref. 62b.

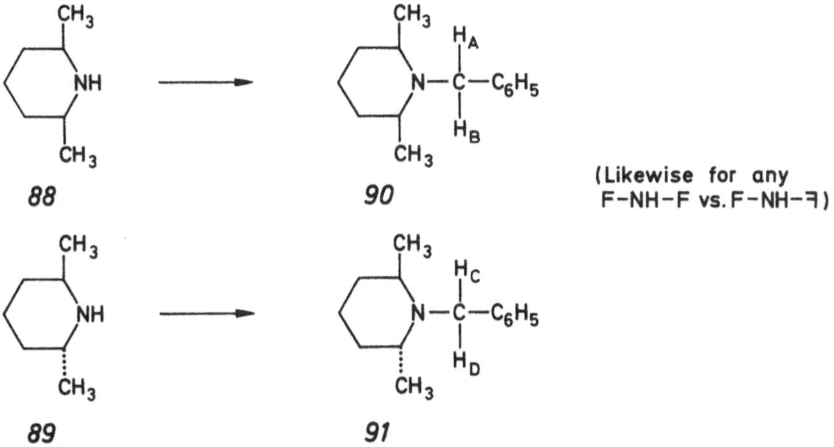

Fig. 36. Achiral probe to distinguish *dl* and *meso* forms

An alternative means for distinction of meso forms and *dl*-pairs [61] is depicted in Fig. 36. Benzylation of the amines *88* and *89* gives the N-benzyl derivatives *90* and *91*. In *90*, derived from the meso isomer *88*, H_A and H_B are enantiotopic and hence isochronous; they constitute a single (A_2) signal. In contrast, in *91*, derived from the active isomer or *dl*-pair *89*, the benzylic protons are diastereotopic and hence anisochronous and constitute an AB system.

Use of this methodology is risky when only one stereoisomer is available. If the benzyl derivative displays a single signal, it is not clear whether one deals with a species of type *90* or whether accidental isochrony is encountered in a species of type *91*. If the latter is the case, the method fails even if both stereoisomers are available. An alternative is shown in Fig. 37 [62]. The amine is converted into its 2,4-dinitrobenzenesulfenyl derivative. In this species the N=S bond has considerable double bond character and rotation around it is slow on the NMR time scale at room temperature; moreover, the structure is such that the N—S—C plane is perpendicular to the C—N—C plane; i.e. the species resembles an allene and displays axial chirality. Derivatization of the meso isomer (*92*) thus gives rise to *two* meso forms, one with the dinitrophenyl group up, the other with the group down (Fig. 37, *93a*, *b*). In each isomer, the methyl groups are internally enantiotopic and thus appear as a sole doublet. However, the methyl groups of *93a* and *93b* are externally diastereotopic and therefore anisochronous; two methyl doublets will thus be generated from the meso form and they will usually be of unequal intensity, since the two diastereomers are normally not formed in equal amounts. In contrast, the *dl* isomer, *94*, will give rise to a single product *95*. In this compound the methyl groups are internally diastereotopic and hence also anisochronous, *but their signals will be equal in intensity*. The meso and *dl* isomers can thus be distinguished by signal intensity measurements on the products *93* and *95*; and this measurement is generally possible even if only one isomer is available to begin with. (A slight uncertainty is introduced by the outside possibility that *93a* and *93b*, though diastereomers, might accidentally be formed in equal amounts.)

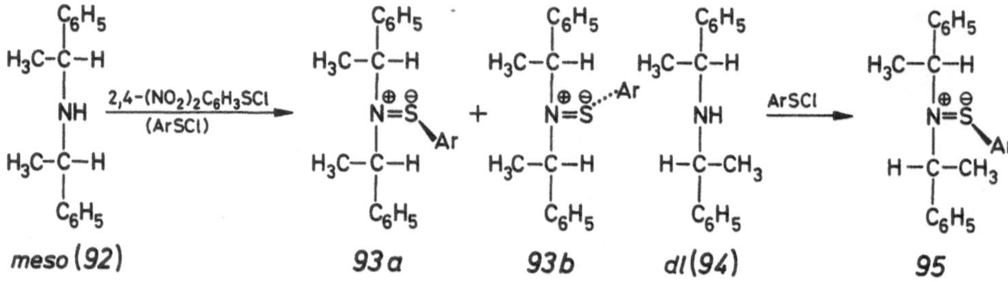

Fig. 37. 2,4-Dinitrobenzenesulfenyl chloride as probe for *dl* and meso forms

We mentioned earlier that the interplay of external and internal diastereotopicity sometimes foils attempts to distinguish *dl* from meso isomers. However, this difficulty is sometimes alleviated inasmuch as internally diastereotopic nuclei may couple with each other whereas externally diastereotopic ones cannot do so. An example [26c] is the distinction of meso- (*96*) and *dl*-2-butylene oxide (*97*) (*cis*- and *trans*-2,3-dimethyloxirane) by means of a chiral shift reagent (Fig. 38). Upon complexation with a chiral shift reagent, the internally enantiotopic C—H protons of the meso isomer *96* become internally diastereotopic and thus anisochronous. The corresponding internally homotopic protons of the active isomer *97* remain homotopic and isochronous. But, in the case of a *dl*-pair, the two enantiomers are converted into diastereomers by complexation with the chiral shift reagent and the protons thus become externally diastereotopic and anisochronous. So far the situation appears stalemated. However, the C—H protons in each enantiomer, because they are isochronous, do not display coupling and the spin system is $X_3AA'X_3'$ or, in close approximation, AX_3: the protons appear as a single quartet. In contrast, the protons of the meso form, being anisochronous do couple and the system is a much more complex X_3ABY_3; the two cases are clearly distinguishable [26c].

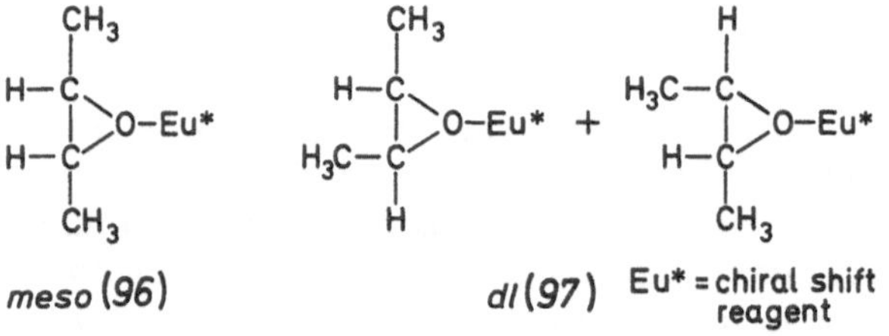

Fig. 38. Distinction of *meso*- and *dl*-2,3-epoxybutane by chiral shift reagent

Use of diastereotopic probes for determination of absolute (as distinct from relative, e.g. meso vs. *dl*) stereochemistry is rare; an example relating to chiral amine oxides is shown in Fig. 39 [26b]. The solute-solvent complex shown, composed of the (*S*)-amine oxide and (*S*)-phenyltrifluoromethylcarbinol, has the ethyl group of the

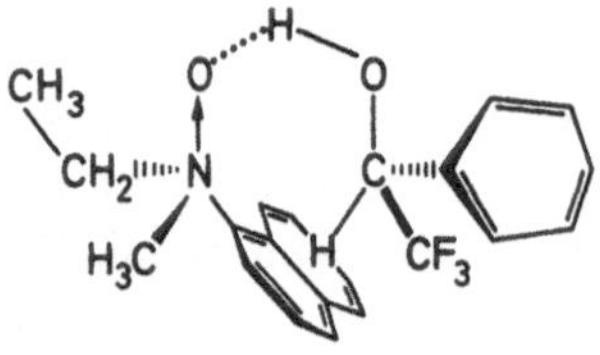

Fig. 39. Determination of absolute stereochemistry through NMR in chiral solvent
[Reprinted with permission from W. H. Pirkle, R. L. Muntz, I. C. Paul, J. Am. Chem. Soc., *93*, 2817 (1971). Copyright 1971, American Chemical Society.]

amine oxide placed in the shielding region of the phenyl moiety of the solvent; the ethyl protons in this diastereomeric complex will therefore resonate upfield of the ethyl protons of the diastereomeric complex from (S)-solvent and (R)-amine oxide. The reverse will be true of the protons of the methyl group of the amine oxide.[16] This method of configurational assignment suffers, of course, from the usual limitation that it is dependent on the correctness of the model on which it is based.

So far in this section we have discussed the use of stereoheterotopic probes in configurational assignment. We now come to the problem of assigning the stereochemical placement (pro-R or pro-S) of the stereoheterotopic groups themselves. One way of doing this involves replacement of the prochiral by a chiral center, for example to replace $RR'CH_2$ by $RR'CHD$ or $RR'CMe_2$ by $RR'C^{12}Me^{13}Me$ or by $RR'C(CH_3)CD_3$. The groups at the chiral center may then be distinguished by the classical methods of configurational determination. Finally they must be correlated with the corresponding groups at the prochiral center.

If the groups in question are enantiotopic, the correlation of the chiral with the prochiral center is in most cases effected through enzymatic reactions. For example, if an enzyme abstracts the deuterium rather than the hydrogen atom from (R)-RR'CHD it will abstract the pro-R rather than the pro-S hydrogen in $RR'CH_2$. Another approach would be to observe a known enantiomer of RR'CHD by NMR in a chiral solvent or in the presence of a chiral shift reagent. If, under these circumstances, the position of the CHD proton is different from that of the corresponding proton in the other enantiomer[17] then the position of this proton, say, in (S)-RR'CHD will correspond, save for small isotope effects, to the position of the pro-R proton in RR'CHH (Fig. 40). A case of this type (except that it involves covalent bond formation) has been described by Mislow and Raban [63a)] and is shown in Fig. 41. It was found that in the (R)-O-methylmandelate of (S)-(+)-2-propanol-1,1,1-d_3 of known configuration [63b)], the (sole) proton doublet of the CH_3 group (A) of the alcohol corresponds to the higher field doublet (A) of the corresponding (R)-O-methylmandelate of unlabeled 2-propanol (Fig. 41, right). The lower-field CH_3-doublet (B) in the unlabeled material disappears in the trideuterated species. It may thus be concluded that the lower-field signal is due to the pro-S methyl group B and the higher-field one to the pro-R group A.

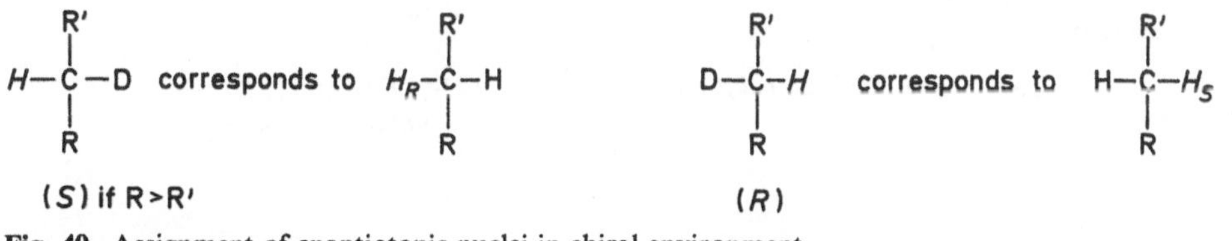

$$H-\underset{\underset{R}{|}}{\overset{\overset{R'}{|}}{C}}-D \text{ corresponds to } H_R-\underset{\underset{R}{|}}{\overset{\overset{R'}{|}}{C}}-H \qquad\qquad D-\underset{\underset{R}{|}}{\overset{\overset{R'}{|}}{C}}-H \text{ corresponds to } H-\underset{\underset{R}{|}}{\overset{\overset{R'}{|}}{C}}-H_S$$

(S) if R>R' (R)

Fig. 40. Assignment of enantiotopic nuclei in chiral environment

---

[16] It is not necessary to have both enantiomers in hand to apply this method. Since the dl-pair displays two sets of signals (of the two diastereomeric complexes), it suffices to compare the dl-pair with one enantiomer.

[17] The obvious way to ascertain this is to look at the racemic mixture under the same experimental conditions and see whether the (now externally diastereotopic) protons of the two enantiomers are distinct. The more convenient way is to make this observation for the internally diastereotopic protons in the unlabeled $RR'CH_2$ in the presence of the chiral shift reagent [52)].

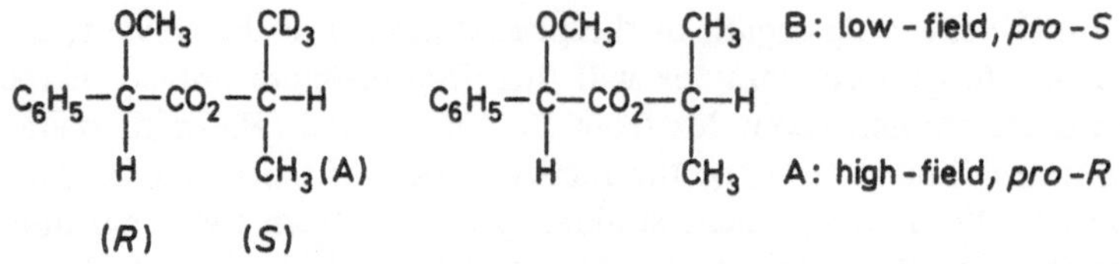

OCH₃ CD₃ — OCH₃ CH₃ structures

Fig. 41. Prochirality assignment of C-methyl groups in isopropyl O-methylmandelate

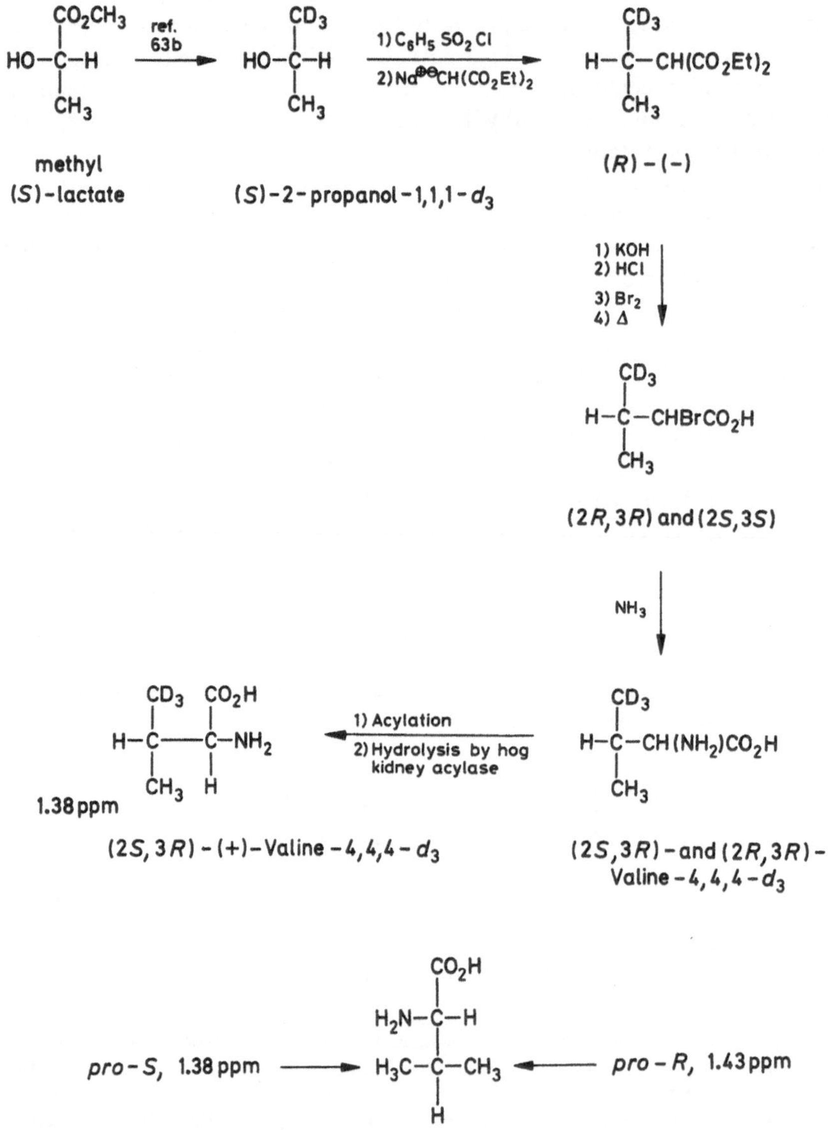

methyl
(S)-lactate

(S)-2-propanol-1,1,1-d_3

(R)-(-)

(2R, 3R) and (2S, 3S)

(2S, 3R)-(+)-Valine-4,4,4-d_3

(2S, 3R)-and (2R, 3R)-
Valine-4,4,4-d_3

1.38 ppm

pro-S, 1.38 ppm ⟶ 98 ⟵ pro-R, 1.43 ppm

Fig. 42. Assignment of diastereotopic methyl groups in L-valine

We have already mentioned that the enantiotopic protons of benzyl alcohol do, in fact, give distinct chemical shifts in the presence of chiral shift reagents [52]. A similar effect can be achieved by "doping" an achiral shift reagent with a chiral complexer [64]. Thus an aqueous solution of sodium or lithium dl-α-hydroxyiso-butyrate, $(CH_3)_2COHCO_2^- M^+$, in the presence of $EuCl_3$ (achiral shift reagent) and L-lactate, $CH_3CHOHCO_2^- M^+$ or D-mandelate, $C_6H_5CHOHCO_2^- M^+$, will display two sets of methyl protons (for the two enantiotopic groups) due to the formation of mixed complexes $(RCO_2)(R'CO_2)_2Eu$. Moreover, assignments of chirality or prochirality can be made by comparing the sense of the shift produced by a given chiral complexer in an unknown situation with that for a compound of known configuration [64]. So far, the method seems to be confined to α-hydroxyacids.

The configurational assignment of the isotopically labeled analog required in all these cases may, of course, be achieved by synthesis from a chiral precursor. A case in point, but relating to diastereotopic nuclei, is shown in Fig. 42 [65a, b]. Valine (*98*) has diastereotopic methyl groups resonating at 1.38 ppm and 1.43 ppm (proton spectrum). In connection with an enzymatic transformation of the molecule, it became of importance to determine which group was which. The methyl groups were introduced stereospecifically starting from (S)-$(+)$-2-propanol-d_3 and the two diastereomers ultimately obtained were separated by resolution (at the conventional chiral center) by enzymatic means. Corresponding assignments with ^{13}C labeled methyl groups have also been described [65c, d].

By an analogous type of reasoning it has been ascertained [66a] which of the two diastereotopic methyl groups at C-12 in Vitamin B-12 (Fig. 43) is derived from methionine: it is the *pro-R* group.

Gerlach [66] has suggested that the absolute configuration of primary alcohols,

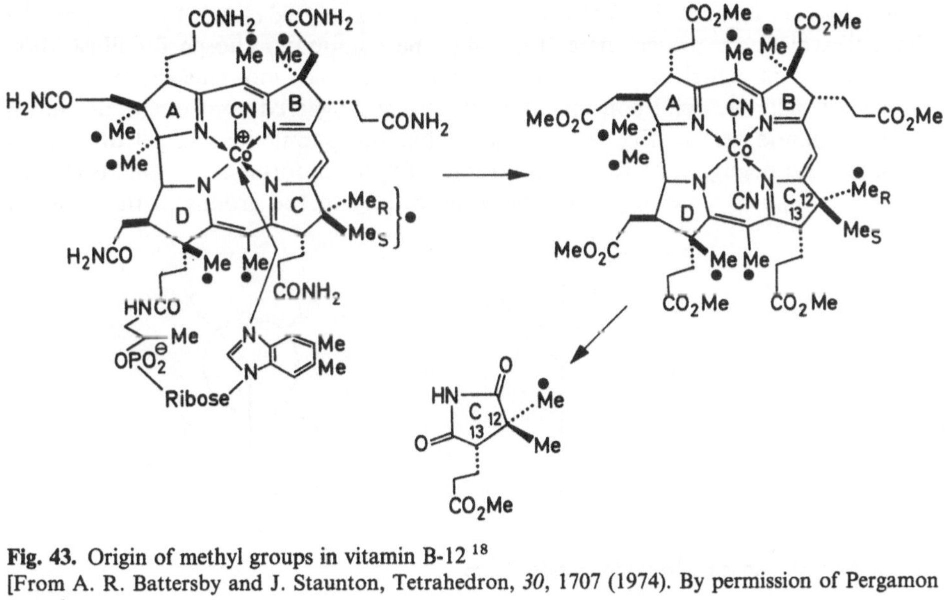

Fig. 43. Origin of methyl groups in vitamin B-12 [18]
[From A. R. Battersby and J. Staunton, Tetrahedron, *30*, 1707 (1974). By permission of Pergamon Press.]

---

[18] The dotted methyl groups are derived from methionine.

RCHDOH, can be determined by the relative chemical shift of the carbinyl protons in the corresponding (—)-camphanate esters in the presence of the (achiral) shift reagent Eu(dpm)$_3$.

An assignment of the chemical shifts of the diastereotopic oxygen nuclei in α-phen-

ethyl phenyl sulfone, $C_6H_5CH(CH_3)\overset{\overset{O}{\|}}{\underset{\underset{O}{\|}}{S}}CH_3$ has been effected by means of ^{17}O NMR

spectroscopy [67]. The sulfones were obtained from the corresponding ^{17}O-labeled, diastereomerically pure (RR*)- and (RS*)-α-phenethyl phenyl sulfoxides by oxidation with unlabeled m-chloroperbenzoic acid. This manner of preparation defines the relative configuration of the α-phenethyl moiety and the labeled sulfone group on the likely assumption that the oxidation proceeds with retention of configuration; the difference in chemical shift between the two ^{17}O nuclei amounts to 4–6 ppm depending somewhat on the solvent. Even larger differences are found in the corresponding α-phenylpropyl analog (6–10 ppm) [67]. By inference, it follows which of the two diastereotopic oxygen atoms (pro-R or pro-S) in a randomly labeled α-phenethyl or α-phenylpropyl phenyl sulfone corresponds to which signal, although, in fact, the signals could not be resolved when the level of ^{17}O was at the natural abundance.

4.3 Origin of Anisochrony

The early history of the anisochrony of diastereotopic groups is turbid because there was uncertainty as to whether the cause for the anisochrony was conformational, intrinsic, or both. The problem was finally analyzed clearly by Gutowsky [23] whose treatment we present here (Fig. 44). The compound chosen for illustration is CxxyCabc in which the x-nuclei are diastereotopic and anisochronous. For simplicity's sake we shall consider only the three staggered conformations shown (for one enantiomer) in Fig. 44, assuming that the populations of all other conformations are negligible.[19] The chemical shift of x_1 in conformers A, B and C may be denoted as $\delta_{a/b}$, $\delta_{a/c}$ and $\delta_{b/c}$ respectively, according to the groups at the adjacent

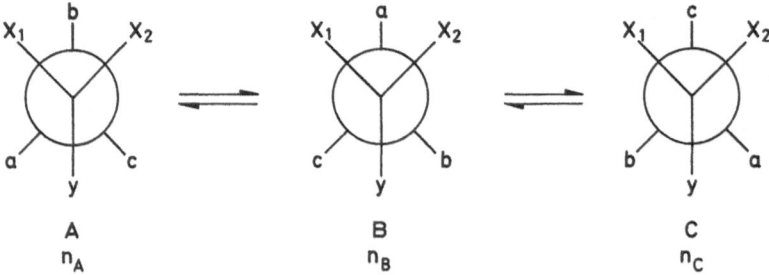

Fig. 44. Anisochronous nuclei (x) in mobile system

---

[19] Strictly speaking this is not correct, since there will be a Boltzmann distribution of molecules among *all* possible conformations [68].

carbon which are gauche to x_1. If n_A, n_B and n_C are the mole fractions of A, B and C respectively, it follows that the average chemical shift of nucleus x_1 is

$$\delta_1 = n_A\delta_{a/b} + n_B\delta_{a/c} + n_C\delta_{b/c} \tag{i}$$

By the same token, the average chemical shift of x_2 is

$$\delta_2 = n_A\delta_{b/c} + n_B\delta_{a/b} + n_C\delta_{a/c} \tag{ii}$$

Inspection of equations (i) and (ii) immediately discloses that since, ordinarily, $n_A \neq n_B \neq n_C$, $\delta_1 \neq \delta_2$, i.e. x_1 and x_2 are anisochronous. It should be noted that, contrary to some misstatements in the literature, this conclusion is independent of the rate of rotation of the CxxyCabc system about the carbon carbon bond which rotation is assumed, throughout, to be fast on the NMR time scale. (See below for what happens in the limit of slow rotation.)

It might then appear, at this point, that the anisochrony was due to the unequal population of the three conformations depicted in Fig. 44. Let us therefore consider the case (hypothetical or otherwise) where $n_A = n_B = n_C$ ($= {}^1/_3$). Inspection of equations (i) and (ii) might, at first glance, imply that, in that case, δ_1 and δ_2 are equal. But, on more careful inspection, it turns out that this is a fallacy spawned by the inadequacy of the notation. The assumption leading to this fallacious result is that $\delta_{a/b}$ in eqn. (i) is the same as $\delta_{a/b}$ in eqn. (ii) (and likewise for $\delta_{b/c}$, $\delta_{a/c}$). In fact, however, the neighboring a of x_1 in A is not the same as the neighboring a of x_1 in B. In the former case, passing beyond a from x_1 one reaches y. In the latter case (B), proceeding from x_2 beyond a one reaches x_1. Hence the environment a/b of x_1 and the shift $\delta_{a/b}$ in A is *not* the same as the environment a/b of x_2 and the shift $\delta_{a/b}$ in B: there is an *intrinsic difference* so that even if $n_A = n_B = n_C$, $\delta_1 \neq \delta_2$. The conclusion, then, is that both the conformation population difference and the intrinsic difference[20] in chemical shift within each conformer contribute to the observed anisochrony of diastereotopic nuclei in conformationally mobile systems [23].

That intrinsic difference is indeed of practical significance was first pointed out by Raban [69] on the basis of data for $BrCF_2CHClBr$ provided by Newmark and Sederholm [70] which are shown in Table 3 on page 38. Conformers A, B and C correspond to the diagrams in Fig. 44 with $x_1 = F_1$, $x_2 = F_2$, y = Br, a = H, b = Br and c = Cl. The gauche neighbors of each fluorine nucleus are indicated in the parentheses in Table 3. It is immediately obvious that not only are F_1 and F_2 anisochronous in each conformer[21] but also there is a substantial difference between nuclei in *apparently* similar environments (*vide supra*). The difference

---

[20] Mislow et al. [30] have pointed out that the distinction between population difference and intrinsic difference is artificial: nuclei are either symmetry related (i.e. interchanged by operation of a symmetry element), in which case they are homotopic or enantiotopic and thus isochronous, or they are not so related, in which case they are diastereotopic or constitutionally heterotopic and therefore anisochronous. While this is certainly correct, the present author believes that the dissection between population and intrinsic difference, like many such dissections in science, is at least pedagogically and possibly in some situations even heuristically useful.

[21] More so in A and C than in B; evidently the halogen environments are similar to each other (Br, Cl) but quite different from that of hydrogen.

Table 3. Low-Temperature-Fluorine NMR Data for
$F_2BrC—CHClBr^a$

Conformer	v_{F_1}		v_{F_2}	
A	2268.4	(H/Br)	3298.8	(Br/Cl)
B	2584.3	(H/Cl)	2628.2	(H/Br)
C	3374.8	(Br/Cl)	2467.4	(H/Cl)
average[b]	2742.5		2798.1	
found[c]	2631.3		2819.1	

[a] In CF_2Cl_2 at 123 K; shifts in Hz from CF_2Cl_2 at 56.4 MHz.
[b] Calculated value. [c] At 303 K, experimental values.

$v_{F_2} - v_{F_1}$ is 359.8 Hz for the H/Br −116.9 Hz for the H/Cl and −76 Hz
for the Br/Cl environments, with the average intrinsic difference being 55.6 Hz.[22]
This is a substantial fraction (nearly a third) of the total observed shift difference
of F_1 and F_2 at room temperature (187.8 Hz); the difference between the two
numbers, 132.2 Hz, is conventionally [69'] ascribed to the population difference of
the three conformers which, in the above case, happens to be relatively small
(n_A = 0.413, n_B = 0.313, n_C = 0.274) [70].

Attempts were also made to establish the existence of an intrinsic shift difference
by measuring the shift difference between diastereotopic nuclei as a function of
temperature. It was argued that as the temperature increases, the population
difference between conformers should vanish and the residual chemical shift
difference (presumably extrapolated to infinite temperature) should be an indicator
of the intrinsic difference. However, it has been pointed out [40] that this assumption
is fraught with complications, stemming from the fact that shifts in individual conformers change with temperature, that the ratio of conformers does not necessarily
converge to unity at high temperature (it will fail to do so when the conformers
differ in entropy) and that the accessible temperature range is often too small for
comfort. Thus if the individual differences in shift are opposite in sign in different conformers (*vide supra*), increase in population of a conformer whose individual shift
has the same sign as the population shift may well lead to an initial increase in overall
shift difference as the temperature is raised rather than a decrease toward the
averaged intrinsic value.

Another elegant way of demonstrating intrinsic non-equivalence, but in a system
at room temperature, was suggested by Mislow and Raban [5] and reduced to
practice by Binsch and Franzen [71] and subsequently by McKenna, McKenna and
Wesby [72]. Two of the molecules studied, a bicyclic trisulfoxide *99* [71] and a
quinuclidine derivative *100* [72] are shown in Fig. 45. In both cases the presence of a
three-fold symmetry axis of one of the ligands (the bicyclic trisulfoxide moiety in *99*,
the quinuclidine moiety in *100*) assures that the three conformers possible by virtue
of rotation about the C—C or N—C bond indicated in heavy type are equally

---

[22] It should be noted that, since the H/Br difference is opposite in sign to the H/Cl and Br/Cl differences,
the average intrinsic shift difference is considerably smaller than the absolute values of the
differences in the individual conformers.

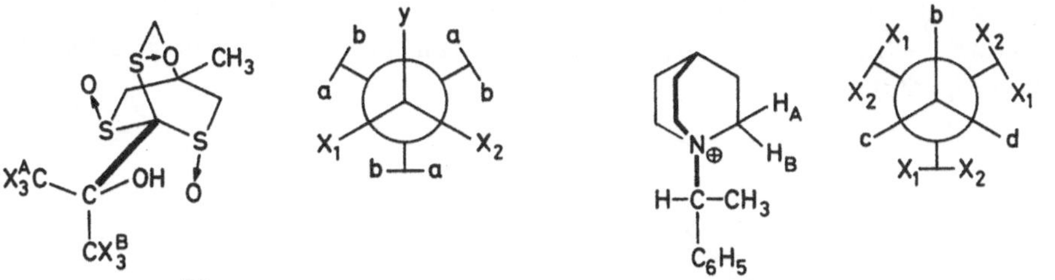

99 [39]   102   100 [40]   101

X = H, δ_{AB} = 0.038 ppm (in C_5H_5N)

X = F, δ_{AB} = 0.282 ppm

δ_{AB} = 0.095 ppm

(decoupled spectrum)

Fig. 45. Molecules displaying intrinsically anisochronous nuclei

populated. The difference in chemical shifts of the CH_3 or CF_3 groups in 99 and the H_A and H_B methylene protons in 100 must therefore be intrinsic in nature. Additional examples of type 100 (general formula 101; the general type formula for 99 is 102) have been adduced [71b, 73, 74].

4.4 Conformationally Mobile Systems

In this section we shall deal briefly with the problem of averaging of heterotopic nuclei. In general, the symmetry properties of a given species are dependent on the time scale of observation in that the symmetry of structures averaged by site or ligand exchanges may be higher than the symmetry in the absence of such exchanges. For the present purpose it is significant that structures lacking C_n or S_n axes may acquire such axes as a result of averaging. It follows that diastereotopic nuclei many become enantiotopic, on the average, through operation of S_n, or they may become equivalent through development of a C_n; in other words, averaging may turn anisochronous nuclei into isochronous ones.

To explore the full potential of what is sometimes called "dynamic NMR" [54], i.e. NMR studies involving site and ligand exchange, is beyond the scope of this chapter and the reader is referred to numerous reviews [40, 41, 54, 75–82]. Only a few examples of the application of this technique can be given here, e.g. in the study of ring inversion, rotation about single bonds and inversion at nitrogen.

An example of ring inversion has already been presented: 1,1-difluorocyclohexane (Fig. 32).

At room temperature the two chair forms average and the average symmetry is that of a planar molecule (C_{2v}) [83] in which the fluorine atoms are related by the C_2 axis and hence equivalent. Thus the room temperature spectrum of 1,1-difluorocyclohexane displays a single (except for proton splitting) chemical shift for the two fluorine atoms, as shown in Fig. 46 [84, 85]. In contrast, at −110 °C the spectrum shows the expected AB pattern for the diastereotopic fluorine atoms expected from the individual structures shown in Fig. 46. As the temperature is gradually raised, the two doublets broaden and merge into two peaks which on further warming eventually coalesce into one single peak at what is called the "coalescence temperature" — in this case —46 °C. (The spectrum just above coalescence is also shown in Fig. 46.)

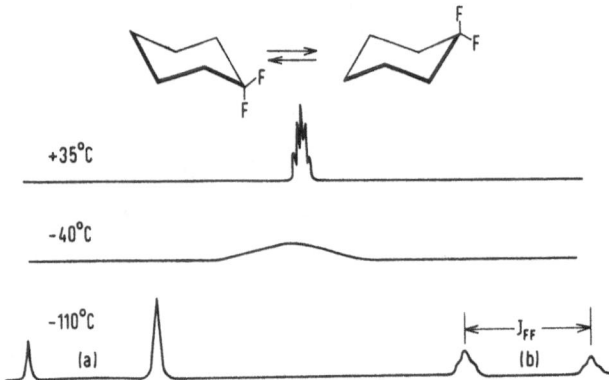

+35°C

−40°C

−110°C

(a) |←— J_{FF} —→| (b)

Fig. 46. ^{19}F NMR signals of 1,1-difluorocyclohexane at various temperatures; signals of **a** equatorial and **b** axial fluorine atoms [From J. D. Roberts, Angew. Chem. Int. Ed. Engl., 2, 53 (1963) by permission of Verlag Chemie.]

An alternative point of view is to recognize that F^1 in structure A (Fig. 32) is externally homotopic with F^2 in structure B (and, similarly, F^2 in A and F^1 in B). Thus F^2 in B has the same shift as F^1 in A (and, likewise, for F^1 in B and F^2 in A); it follows that when the interchange of A and B is rapid on the NMR time scale, all four fluorine nuclei have the same shift, i.e. become isochronous and do not couple with each other.

A simple formula for determining the rate of site exchange between two equally populated sites is

$$k_{coal} = {}^1/_2 \pi \, \Delta v \sqrt{2} = 2.221 \, \Delta v \quad {}^{54, 86, 87)} \tag{iii}$$

This equation is valid only for a (non-coupled) single site-exchanging nucleus, for example the proton in cyclohexane-d_{11} (observed with deuterium decoupling). In the case of geminal exchanging nuclei, as in 1,1-difluorocyclohexane (proton-decoupled), the equation

$$k_{coal} = {}^1/_2 \pi \sqrt{2} \sqrt{(v_1 - v_2)^2 + 6J^2} \quad {}^{54, 87, 88)} \tag{iv}$$

should be used, J being the coupling constant of the two nuclei in question.[23] Both v and J are measured at temperatures much below the coalescence temperature. For 1,1-difluorocyclohexane [84] $v_1 = 1522$ Hz, $v_2 = 638$ Hz, $J = 237$ Hz and thus k = 1058 sec^{-1} at the coalescence temperature of —46 °C. From this information,

---

[23] There has been a controversy [54, 89] regarding the accuracy of equations (iii) and (iv), i.e. the validity of inferring rate constants from coalescence temperatures. It now appears [89] that, provided no additional coupling is present, and provided $\Delta v > 3$ Hz (eqn. (iii)) and $\Delta v > J$ (eqn. (iv)), the simple coalescence method gives rate constants within 25% of those obtained by more sophisticated line shape analysis *(vide infra)*. This leads to an error in $\Delta G°$ no greater than that produced by the uncertainty of temperature measurement (± 2 °C [89]) in the NMR probe.

in turn, one can calculate the free energy of activation for the site exchange from the formula

$$k = (\varkappa k_B T/h)\, e^{-\Delta G^{\neq}/RT} = (\varkappa k_B T/h)\, e^{-\Delta H^{\neq}/RT}\, e^{\Delta S^{\neq}/RT} \quad \text{54, 86)}$$

(v)

where k is the rate constant for site exchange, $\varkappa$ is the transmission coefficient usually taken as unity), k_B is Boltzmann's constant, h is Planck's constant, T is the coalescence temperature, $\Delta G^{\neq}$ is the free energy of activation, $\Delta H^{\neq}$ is the enthalpy of activation and $\Delta S^{\neq}$ is the entropy of activation. From the data for 1,1-difluoro-cyclohexane, $\Delta G^{\neq} = 9.71$ kcal/mol [84].

It should be noted that in exchanges between two unequally populated sites, for example in 4-chloro-1-protiocyclohexane-d_{10} or 4-chloro-1,1-difluorocyclohexane, the rate constants calculated from eqns. (iii) and (iv), respectively, are the average of the forward and reverse rate constants, which we shall design as k_A and k_B, respectively. These constants are, of course, no longer equal when the populations of A and B in $A \underset{k_B}{\overset{k_A}{\rightleftharpoons}} B$ are unequal at equilibrium, since $k_A/k_B = K$, the equilibrium constant. In this case an approximation formula [90] (vi) may be used; k may be taken as the average rate constant calculated by formulas (iii) and (iv)[24] and Δn is the difference in mole fractions of the two species A and B at equilibrium (i.e. $\Delta_n = = n_B - n_A$, assuming B is the predominant species).

$$k_A = (1 + \Delta n)\, k \quad \text{and} \quad k_B = (1 - \Delta n)\, k \quad [24]$$

(vi)

A more general method of measuring site exchange rates is the method of line shape analysis [54, 87]. In this method one compares the shape of the broadened lines some ten or twenty degrees above and below the coalescence temperature (as well as in the fast and slow exchange limit) with the line shape computed by means of formulas which include the rate of exchange. This method permits determination of k over a range of temperatures and thus — through a plot of $\Delta G^{\neq}$ vs. $1/T$ — of $\Delta H^{\neq}$ and $\Delta S^{\neq}$ (though the accuracy of determining $\Delta H^{\neq}$ and $\Delta S^{\neq}$ is often low). The method is applicable to relatively complex spin systems, not just to singlet or AB exchange.[25] It is considered to be the method of choice in determination of rate constants by NMR. A typical comparison of experimental and computed line shapes, referring to the site exchange in furfural is shown in Fig. 47 [91].

---

[24] This is somewhat different from the treatment in the original reference [90]. Unfortunately, formula (iii) no longer holds strictly when $k_A \neq k_B$, but the approximate treatment given here gives errors of less than 25% (cf. footnote 23) for mole fractions between 0.2 and 0.8. Outside of this region — or for more accurate results inside the region — the original graphic treatment [90] should be used.

[25] Because each chemical shift difference between exchanging sites as well as each spin coupling constant gives rise to a coalescence of its own when Δv or $J \approx k$ (where k is the rate of site exchange), a system having many such parameters will, because of the presence of a multitude of "internal clocks" be more sensitive in the response of its NMR spectrum to temperature changes [87]. Thus, within the limits of feasibility of computer treatment, the more shifts and coupled spins, the better.

It is convenient to have terms for structures such as those in Fig. 32 and 47 (*103*) which differ only in the position of designated nuclei, and for the process of exchange of such heterotopic nuclei. The term "topomers" has been proposed for the interconverting structures, "topomerization" being the process of interchange [92]. An older term, "degenerate isomerization", seems inappropriate since the two structures shown in Fig. 32 are not isomers. "Automerization" has also been used [92]; it properly denotes the identity of the two interconverting structures but does not address itself to the significance of the process of their interconversion.

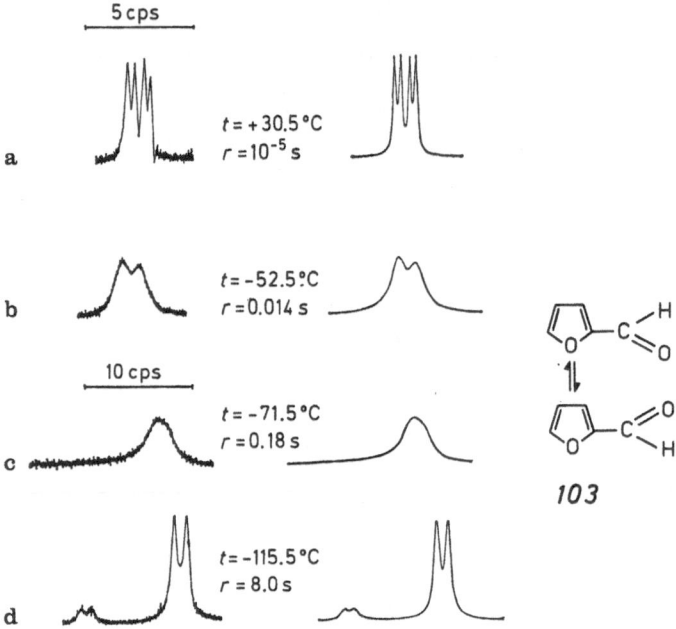

Fig. 47. Experimental and calculated DNMR spectra for the aldehyde proton of 2-furaldehyde [91]. [Reprinted with permission from K. I. Dahlqvist and S. Forsén, J. Phys. Chem., *69*, 4062 (1965) Copyright 1965, American Chemical Society.

We have already referred to studies of rotational isomerism by DNMR. Most barriers to rotation are below 5 kcal/mol and cannot be investigated by NMR but there are a number of cases where the barrier *is* high enough for NMR study [94]. Frequently, such cases relate to partial double bonds [95, 96]. Even homotopic ligands such as the fluorine nuclei in a CF_3 group or the methyl groups (protons or ^{13}C signals) in a *t*-butyl group may become diastereotopic in the slow exchange limit; examples [94] are shown in Fig. 48.

As the last example of coalescence of signals of diastereotopic nuclei due to site exchange in solution we present the case of the Schiff base of α-naphthyl isopropyl ketone and benzylamine (*106*, Fig. 49) [99]. This molecule possesses a prochiral axis by virtue of the fact that coplanarity of the C=N double bond with the naphthalene ring leads to strong steric interactions; the coplanar conformation

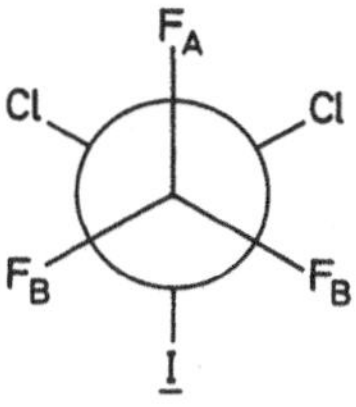

Fig. 48. Diastereotopic fluorines in CF_3 and methyls in $C(CH_3)_3$

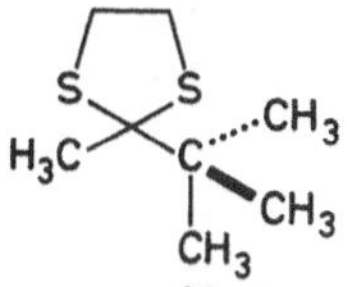

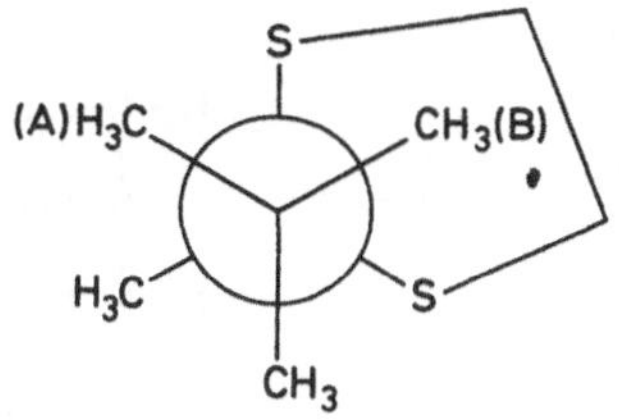

thus corresponds to a maximum between enantiomeric, chiral, non-coplanar structures. In such structures, both the protons of the benzyl ($PhCH_2N$) group and the methyl ligands of the isopropyl group are diastereotopic, and, indeed, in most solvents these groups are anisochronous in the proton spectrum and the methyl groups are also anisochronous in the ^{13}C spectrum. Upon heating the solution in appropriately high-boiling solvents to 150 °C, the anisochrony abruptly disappears (i.e. the signals collapse) presumably because rotation about the Ar—C bond, heretofore hindered, becomes rapid on the NMR time scale. Interestingly enough, anisochrony also disappears in solvent carbon tetrachloride at room temperature, but reappears not only on lowering the temperature (to —20 °C) but also on raising it (to 43 °C). The disappearance of the anisochrony at room temperature is therefore an accidental one; at room temperature the two methyl-H signals have coincidentally the same chemical shift, but the shift of the one group increases and that of the other decreases with temperature, so that at higher or lower temperatures than ambient, the signals will diverge (in opposite directions). No such phenomenon is seen in the ^{13}C spectrum where the methyl groups remain anisochronous from —20 up to nearly 150 °C.

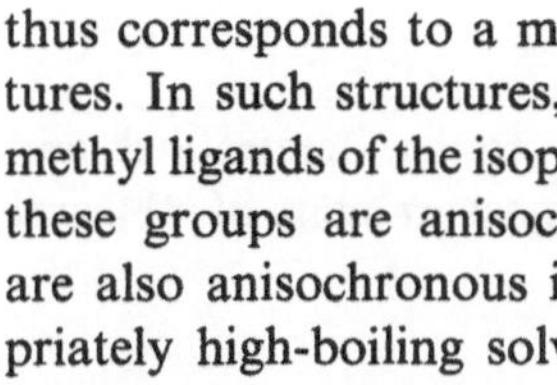

Fig. 49. Prochirality in N-[2-methyl-1-(1-naphthyl)propylidene]-benzylamine

We conclude this section with a discussion of inversion of amines of the type $NR_1R_2R_3$ [75,76]. In general this process is too rapid to be studied except in special circumstances [100]. Nevertheless, Saunders and Yamada [101] were able to determine the very high rate of inversion of dibenzylmethylamine (Fig. 50) ($k = 2 \times 10^5 \text{ sec}^{-1}$ at 25 °C) by the elegant trick of partially neutralizing the amine with hydrochloric acid. Since the hydrochloride cannot invert, the benzylic protons in it are diastereotopic and hence anisochronous. Only the small amount of free amine in equilibrium with the salt at a given pH (the measurements were carried out on the acid side) inverts at the rate indicated and it can be easily shown [101] that $k_{obs.} = k \cdot [\text{amine}]/[\text{salt} + \text{amine}]$ where $k_{obs.}$ is the observed rate of site exchange of the diastereotopic protons at

43

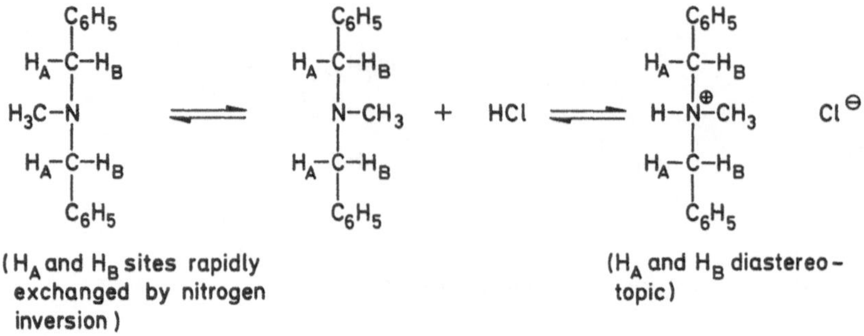

$$
\begin{array}{ccc}
\underset{\text{(H}_A \text{ and H}_B \text{ sites rapidly}}{
\begin{array}{c}
C_6H_5 \\
H_A-C-H_B \\
| \\
H_3C-N \\
| \\
H_A-C-H_B \\
| \\
C_6H_5
\end{array}}
&
\rightleftharpoons
&
\begin{array}{c}
C_6H_5 \\
H_A-C-H_B \\
| \\
N-CH_3 \\
| \\
H_A-C-H_B \\
| \\
C_6H_5
\end{array}
+ HCl \rightleftharpoons
\begin{array}{c}
C_6H_5 \\
H_A-C-H_B \\
| \\
H-N^{\oplus}-CH_3 \\
| \\
H_A-C-H_B \\
| \\
C_6H_5
\end{array}
\quad Cl^{\ominus}
\end{array}
$$

(H$_A$ and H$_B$ sites rapidly exchanged by nitrogen inversion)

(H$_A$ and H$_B$ diastereotopic)

Fig. 50. Inversion of dibenzylmethylamine

a given pH, k is the rate constant for amine inversion to be determined and the quantity in the fraction can be ascertained from the measurement of pH and the known basicity of the amine.

4.5 Spin Coupling Non-Equivalence (Anisogamy)

Nuclei which are anisochronous in general not only couple with each other but also differ in coupling constants with respect to a third nucleus. Such non-equivalence with respect to spin coupling may be called "anisogamy" and the nuclei are "anisogamous" [102] as well as anisochronous. Thus the $^{13}C-{}^1H$ spin coupling constants of the diastereotopic methylene protons in $CH_3CH(OCH_2CH_3)_2$ are 139.6 and 141.0 Hz [103] and the diastereotopic methyl protons in $C_6H_5P(CH(CH_3)_2)_2$ have $^{31}P-{}^1H$ coupling constants of 11.0 and 14.7 Hz [104].

However, even isochronous nuclei may be anisogamous. Thus, whereas the two aromatic protons in 1,3-dibromo-2,5-difluorobenzene, *107* (Fig. 51) are both isochronous and isogamous and thus give rise to a single resonance, the two protons ortho to bromine in *p*-chlorobromobenzene (*108*, Fig. 51) are isochronous but not isogamous; the same is true of the two protons ortho to chlorine. Inspection of *108* shows that H$_A$ will be differently coupled to H$_B$ (to which it is constitutionally heterotopic) than H$_C$ is to H$_B$; therefore $J_{AB} \neq J_{CB}$, i.e. H$_A$ and H$_C$ are anisogamous. The system is of the AA'BB' type [105]. While this subject is discussed in all textbooks on nuclear magnetic resonance, it is worth while to state here the symmetry criterion for isogamy (or anisogamy) [106]: If in a spin system of the type A_2B_2 ... (or AA'BB' ...) substitution of one of the B's by a different nucleus Z leads to a system A_2BZ in

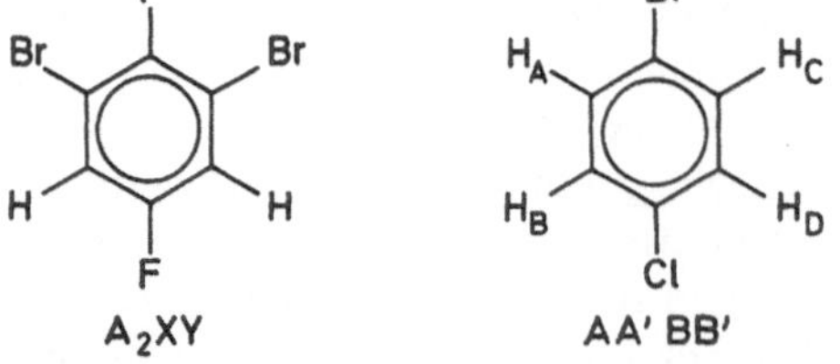

A$_2$XY

107

AA' BB'

108

Fig. 51. Isogamous and anisogamous nuclei

which the remaining homomorphic nuclei A are no longer interconvertible by a symmetry operation[26] (C_n or S_n, including σ) then B and B' are anisogamous. But if the A nuclei remain interconvertible by a symmetry operation, then the B nuclei are isogamous. By this criterion the protons in CH_2F_2 and $CH_2=C=CF_2$ are isogamous since they remain enantiotopic, and hence isochronous, in CH_2FBr and $CH_2=C=CFBr$ (this would be true even if Br were a palpably magnetic nucleus). But in $CHF=C=CHF$ or cis- or trans-$CHF=CHF$ the protons are anisogamous, for whereas they are related by a C_2 axis in the compounds shown (and hence are isochronous), they are no longer so related in $CHF=C=CHBr$ or cis- or trans-$CHF=CHBr$. The same is true for the methylene protons in $BrCH_2CH_2CO_2H$ (they become diastereotopic in $BrCH_2CHClCO_2H$ or $BrCHClCH_2CO_2H$).

It is of interest that the protons (or fluorine nuclei) in CH_2F_2 should become anisogamous in a chiral solvent [106], since the enantiotopic protons in CH_2CFBr would become diastereotopic in such a solvent; however, attempts to demonstrate such anisogamy have not so far been successful [106].

5 Prostereoisomerism in Enzyme-catalyzed Reactions

5.1 Prostereoisomerism and Asymmetric Synthesis

We have seen in Section 3 that replacement of stereoheterotopic groups or addition to stereoheterotopic faces gives rise to stereoisomers. The rates of such replacements of one or other of two ligands or additions to one or other of two faces are frequently not the same. In particular, replacements of diastereotopic ligands or additions to diastereotopic faces usually proceed at different rates because the transition states for such replacements or additions are diastereomeric and therefore unequal in energy. For example, the reactions shown in Figs. 18 and 19 not only give rise to diastereomeric products, depending on which ligand or face is involved, but they give these products in unequal, sometimes quite unequal, amounts. Thus reactions of this type display diastereoselectivity [107] or "diastereo-differentiation" [108]. Replacement of enantiotopic ligands or addition to enantiotopic faces gives rise to enantiomeric products, but here replacement of the two ligands or addition to the two faces ordinarily occurs at the same rate, because the pertinent transition states are enantiomeric and therefore equal in energy. This situation changes, however, when the reagent (or other entity participating in the transition state, such as the solvent or a catalyst) is chiral. In that circumstance, the two transition states will, once again, become diastereomeric and the two enantiomeric products will be formed at unequal rates and in unequal amounts: the reaction will be enantioselective [107] or "enantio-differentiating" [108]. In this case, where prochiral starting materials give rise to

---

[26] This criterion should be applied to the conformation of highest symmetry. It does not apply to cases where the A and B nuclei are themselves symmetry equivalent, such as $ClCH_2CH_2Cl$ or p-dichlorobenzene.

chiral products, one speaks of an asymmetric synthesis [28, 29]. While this topic is outside of the scope of the present review, we shall discuss here the question as to which of two stereoheterotopic ligands is displaced in a given reaction. (The complementary problem as to which of two stereoheterotopic faces of a double bond is approached by a reagent in an asymmetric addition reaction is usually trivial, since the stereochemistry of the product will, in almost all cases, reveal the answer to this question. Exceptions occur only when the moiety added to the double bond is identical to an already attached substituent — for example in the reduction of acetaldehyde, $CH_3CH=O$ by hydride to ethyl alcohol, CH_3CHHOH — and in that case, as shown in the next Section, the problem is solved by the use of isotopically substituted reagents or substrates, e.g. the reduction of $CH_3CH=O$ by deuteride or that of $CH_3CD=O$ by hydride.)

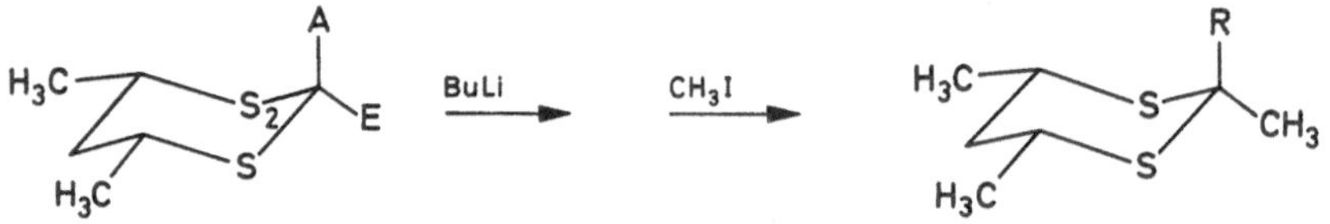

109 a A=E=H
109 b E=H, A=D
109 c E=D, A=H

110 a R=H
110 b R=D

Fig. 52. Steric course of lithiation of conformationally locked 1,3-dithiane

A typical illustration of the problem is provided by the lithiation, followed by electrophilic substitution, of a conformationally locked 1,3-dithiane 109a, shown in Fig. 52, to give exclusively the equatorial alkylation product 110a [109]. A priori this reaction may proceed stereospecifically with abstraction of an equatorial hydrogen (E) followed by substition of the resulting carbanion with retention; or it might proceed stereospecifically by abstraction of the axial hydrogen (A) at C(2) followed by electrophilic substitution with inversion; or, finally, it might involve abstraction of either hydrogen (A or E) to form a carbanion which would then be alkylated selectively from the equatorial side; this reaction course would involve stereoconvergence. To distinguish among the three eventualities, the reaction was carried out after labeling of one or other of the diastereotopic hydrogens at C(2) through substition by deuterium (109b, 109c, Fig. 52). It was thus found that either hydrogen was abstracted by the base (BuLi), though abstraction of the equatorial proton (after correction for isotope effects) was about 9 times faster than that of the axial. However, since the product (110) had over 99% equatorial methyl substitution regardless of the stereochemistry of the starting material at C(2) (109a, 109b, 109c), it must be concluded that the two-step reaction is stereoconvergent [109].

Another, important, application of the use of isotopic labeling to decide which of two stereoheterotopic ligands is involved in a classical chemical reaction is in the work of J. Sicher and his school concerning the mechanism of onium ion elimination [110]. As the result of extensive studies, it was concluded (Fig. 53) that elimination of the *syn*-hydrogen[27] leads to a trans-olefin and that of the *anti*-hydrogen[27] to a cis-olefin (Fig. 53) [110].

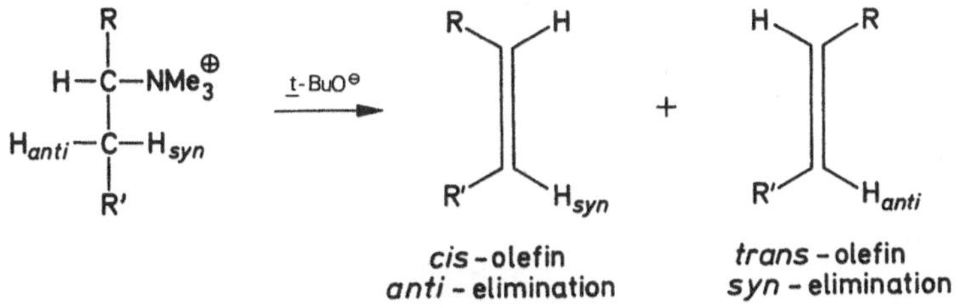

Fig. 53. Stereochemistry of E_2 elimination

By far the most extensive applications of this technique, however — i.e. of ascertaining which of two stereoheterotopic ligands or faces is implicated in a given reaction — have been in enzyme chemistry and the next section will deal with this topic.

5.2 Applications to Enzyme-catalyzed Reactions

Once one finds out which of two stereoheterotopic ligands or faces of a substrate is involved in an enzyme-catalyzed reaction, one is in a position to make a meaningful statement as to the location of the substrate in relation to the active site of the enzyme. While considerations of prostereoisomerism are thus useful in helping elucidate the enzyme-substrate relationship in the activated complex of an enzyme-mediated reaction, it must also be stressed that such considerations in themselves are insufficient to provide the complete picture and that they must necessarily be supplemented by many other techniques in enzyme chemistry.

The literature in the area of prostereoisomerism in enzyme reactions is vast and we must confine ourselves in this section to the discussion of a few representative examples. For more detailed information the reader is referred to a number of review articles [19,32,111-118] and two books [119,120] which have appeared in the last dozen years.

We shall start the discussion with a classical experiment related to the stereochemistry of oxidation of ethanol and reduction of acetaldehyde mediated by the enzyme yeast alcohol dehydrogenase in the presence of the oxidized (NAD$^+$) and reduced (NADH) forms, respectively, of the coenzyme nicotinamide adenine dinucleotide (Fig. 54). The stereochemically interesting feature of this reaction stems from the fact that the methylene hydrogens in CH_3CH_2OH and the faces of the carbonyl in $CH_3CH=O$ are enantiotopic. The question thus arises which of the CH_2-hydrogens

---

[27] The designations *syn* and *anti* are those of the original authors [110] and undoubtedly derive from the fact that the initial work related to ring compounds where the meaning of syn and anti is unequivocal. In the acyclic systems (Fig. 53) the designations refer to a Fischer projection with the alkyl groups at the top and bottom, i.e. eclipsed in the same way as they would be in a small ring.

is removed in the ρxidation and to which of the $C=O$ faces the hydrogen attaches itself in the reduction in the presence of the enzyme and coenzyme.

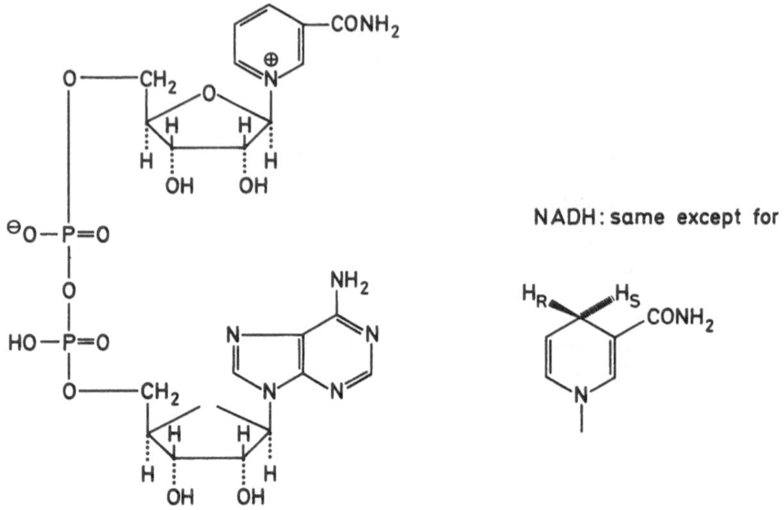

NADH: same except for

Fig. 54. Nicotinamide adenine dinucleotide (NAD^+)

Westheimer, Vennesland and Loewus in 1953 found [121] that reduction of ethanal-1-*d* with NADH in the presence of yeast alcohol dehydrogenase gave ethanol-1-*d* which, upon enzymatic reoxidation by NAD^+, returned ethanal-1-*d* without loss of deuterium. There is thus a "stereochemical memory effect" involved in this reaction: the H and D of the CH_3CHDOH do not get scrambled but the same H which is introduced in the reduction is the one removed in the oxidation. In the light of what we now know, this is, of course, not surprising since the two methylene hydrogens are distinguishable by bearing an enantiotopic relationship.

When the configuration of the ethanol-1-*d* is inverted by conversion to the tosylate followed by treatment with hydroxide and the inverted ethanol-1-*d* is then oxidized with yeast alcohol dehydrogenase and NAD^+, the deuterium (which has taken the stereochemical position of the original hydrogen) is now removed and the product is unlabeled $CH_3CH=O$.[28] The sequence of events [121] is summarized in Fig. 55.

Later experiments on a larger scale [122] established that the ethanol-1-*d* obtained from $CH_3CD=O$ and NADH (upper right in Fig. 55) was levorotatory, $[\alpha]_D^{28}$ $- 0.28 \pm 0.03$ and this finding, coupled with the elucidation of configuration of (—)-ethanol-1-*d* as S [111] leads to the stereochemical picture summarized in Fig. 55. It follows therefore that the hydrogen transferred from the NADH in the enzymatic reduction attaches itself to the *Re* face of the aldehyde and that this hydrogen thus

---

[28] The sequence is not entirely clean in that some CH_3CDO is also obtained. Probably this is due to incomplete inversion in the tosylate — hydroxide reaction resulting from O—S cleavage (with retention).

becomes H_R in the ethanol; it is H_R (the *pro-R* carbinol hydrogen), in turn, which is abstracted by NAD^+ in the oxidative step.[29]

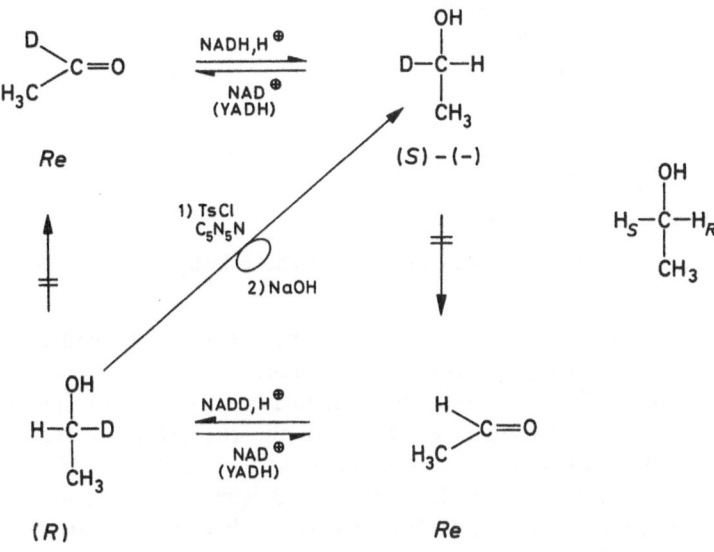

Fig. 55. Oxidation of ethanol and reduction of acetaldehyde by $NAD^+/NADH$ in the presence of yeast alcohol dehydrogenase (YADH)

It is clear that ethanol (and acetaldehyde) must fit into the active site of yeast alcohol dehydrogenase in such a way as to conform to these stereochemical findings. A model for the reduction of a very similar substrate, pyruvic acid (which is reduced by NADH in the presence of liver alcohol dehydrogenase to (*S*)-lactic acid) is shown in Fig. 56 [124]. Here we can discern Ogston's picture of the three-point contact (cf. Fig. 8), one contact being established by the salt bond pyruvate — arginine-H^+, the second by the hydrogen bond (histidine) $N-H \ldots O=C$(pyruvate) and the third one involving delivery of the hydrogen of NADH (bound to the enzyme) to the *Re* face of the $C=O$ of pyruvate.[30] The reduction of acetaldehyde is probably similar though the absence of the COO^- group requires the contact at the third site to be established in a different manner. It is not certain that covalent or ionic bonding is actually involved in this contact; the shape of the enzyme cavity itself (and the attendant hydrophilic and hydrophobic interactions between certain parts of the enzyme and parts of the substrate) may contribute to the required orientation of the substrate.

---

[29] Of course, in the enzymatic oxidation of unlabeled ethanol one cannot operationally discern that the hydrogen abstracted is *pro-R*. Reactions of this type which are "prostereoselective" have been called "stereochemically cryptic" [123].

[30] The same model indicates, of course, that only (*S*)-lactate [not (*R*)-lactate] is formed in the reduction. By the same token it explains why, in the reverse reaction, the enzyme is substrate stereoselective for (*S*)-lactate: (*R*)-lactate, if locked into the enzyme cavity, would have CH_3 rather than $C-H$ juxtaposed with the NAD^+ and could thus not be oxidized.

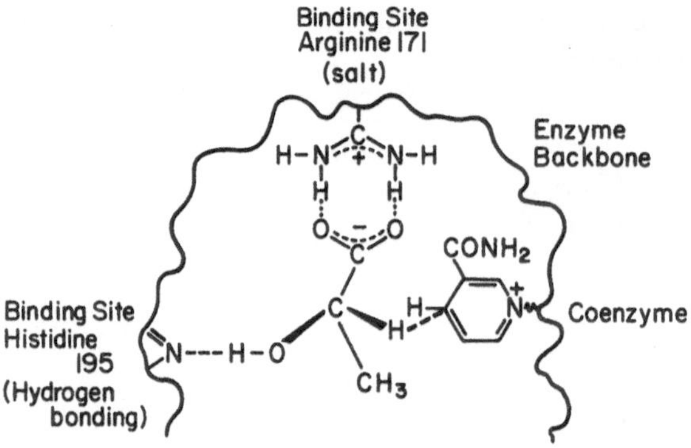

Fig. 56. Reduction of pyruvate by NADH in the presence of liver alcohol dehydrogenase

The study of the stereochemistry of ethanol oxidation and acetaldehyde reduction and the information relating to the topography of the enzyme derived from this study are typical of a large number of other investigations of this type. We wish to point out here that the transfer of the hydrogen to and from the coenzyme involves a stereochemical problem of its own (Fig. 57): In the reductive step, is it H_R or H_S of the dihydronicotinamide moiety which is transferred from the coenzyme to the substrate; correspondingly, in the oxidation step, is the hydrogen abstracted from the substrate added to the *Re* or *Si* face (or, using Rose's nomenclature[38],

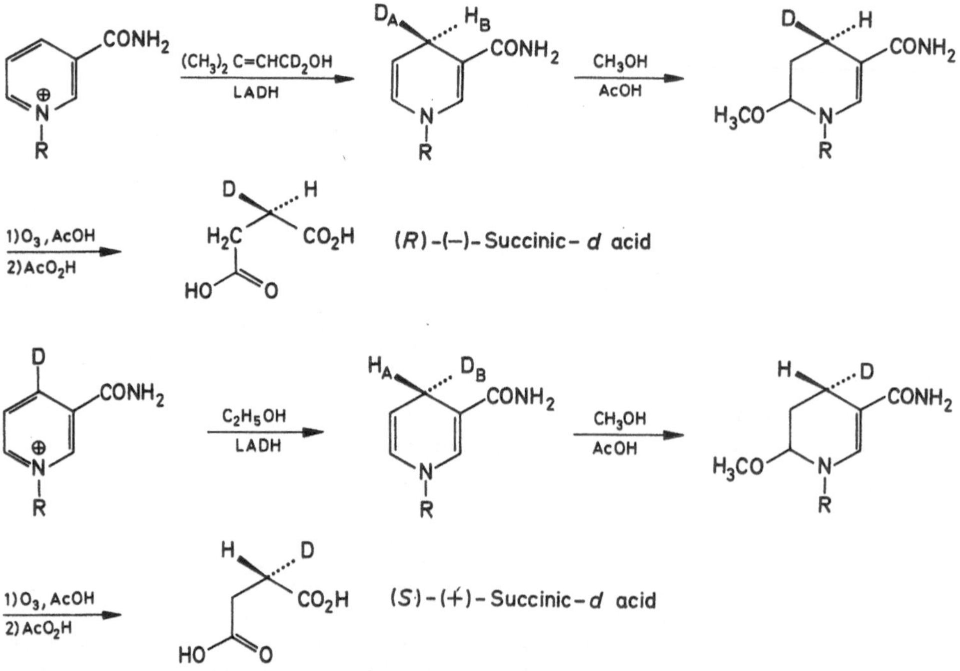

Fig. 57. Prostereoisomerism of hydrogen transferred to C(4) of NADH from alcohol in the presence of liver alcohol dehydrogenase

the β or α face) of the pyridinium moiety of the coenzyme? This question was answered [125] as summarized in Fig. 57.

Deuterium was transferred from a dideuterated alcohol (RCD_2OH) to NAD^+ in the presence of liver alcohol dehydrogenase. This creates a chiral center at C(4) in the NAD^2H formed. Degradation of this material in the manner shown (Fig. 57) yielded (R)-(−)-succinic-d acid recognized by its known ORD spectrum. It follows that the configuration of the NAD^2H formed was R and it is therefore H_R which is transferred from (ant to) the alcohol; attachement of hydride to NAD^+ thus occurs from the Re (β) face.[31] To confirm this finding and to avoid any remote possibility that the β,β-dimethylallyl-d_2 alcohol used as the source of deuterium would behave differently from ethanol, the experiment was repeated with NAD-4-d^+ and ethanol, as shown in Fig. 57 (bottom part). In this case, of course, the ultimate degradation product is (S)-(+)-succinic-d acid.

We next take up the stereochemistry of an enzymatic addition to a C=C double bond: the hydration of fumaric to (S)-malic [127] and the amination of fumaric to (S)-aspartic acid [128]. (Both reactions are of industrial importance [129].) These reactions are summarized in Fig. 58. The absolute configurations of both (−)-malic and (−)-aspartic acids are well known and erythro- and threo-malic-3-d acids have been identified by NMR spectroscopy (being diastereomers, they differ in NMR spectrum) and their configuration has been unambiguously assigned by a synthesis of controlled stereochemistry (Fig. 59) [130, 131]. In the dianion of this acid in D_2O solution, the carboxylate groups are anti (because of electrostatic repulsion) and it follows that the hydrogen atoms are gauche; the coupling constant of these protons is therefore small (J = 4 Hz). In contrast the erythro isomer obtained by biosynthesis (see below) has its hydrogen atoms anti to each other and their coupling constant is thus larger (J = 6–7 Hz).

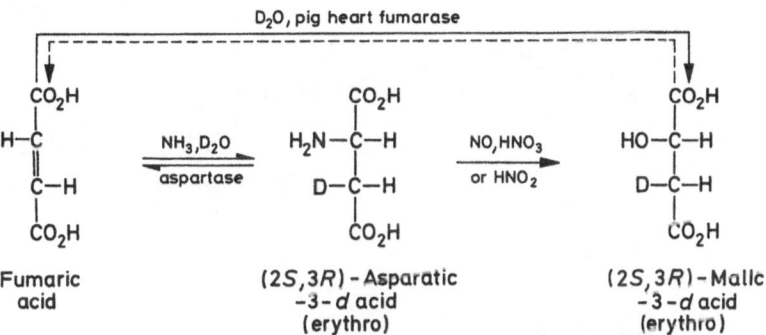

Fig. 58. Stereochemistry of fumarase and aspartase reactions

The fact that, though addition of D_2O to fumaric acid is reversible, it yields only a monodeuterated malic acid and involves recovery of undeuterated fumaric acid only, indicates that the addition and elimination steps are stereospecific and that they proceed with the same stereochemistry. The formation of the erythro isomer of

---

[31] This result is not general; i.e. it does not apply to all oxidation-reduction reactions mediated by NAD^+/NADH [126].

(S)-(—)-malic-3-d acid in the fumarase mediated D$_2$O addition (Fig. 58) relates the absolute stereochemistry of C(3) to that at C(2) and proves that the configuration at C(3) is R. Since the (S)-(—)-aspartic-3-d acid formed in aspartase mediated ammonia addition to fumaric acid (Fig. 58) is converted to the same (2S,3R)-(—)-malic-3-d acid by nitrous acid deamination, and since this reaction does not affect the stereochemistry at C(3), the aspartic acid must also be 3R. (The stereochemistry of the nitrous acid deamination at C(2) involves retention, and this was already well known in the literature.) It follows, then, that the hydrogen added (or abstracted, in the reverse reaction) at C(3) in the conversion of fumaric to malic or aspartic acid is the *pro-R* hydrogen and that attack on the fumaric acid of the proton from either water or ammonia proceeds from the *Re* face at C(3). (This is the front face in Fig. 58.) On the other hand, since the configuration at C(2) in both cases is S, the attack at C(2) in both cases must be from the rear face in Fig. 58, i.e. from the *Si* face. The overall picture, then, is one of anti addition giving the 2S,3R isomer (Fig. 60). Analogous stereochemistry is observed in the addition of water to maleic, citraconic (α-methylmaleic) and mesaconic (α-methylfumaric) acids [111].

We finally return to the subject of citric acid. We posed the question in Section 2 as to whether the four hydrogen atoms in citric acid (Fig. 4) were distinguishable and we have already seen that H$_A$ and H$_B$ (or H$_C$ and H$_D$) are diastereotopic and therefore,

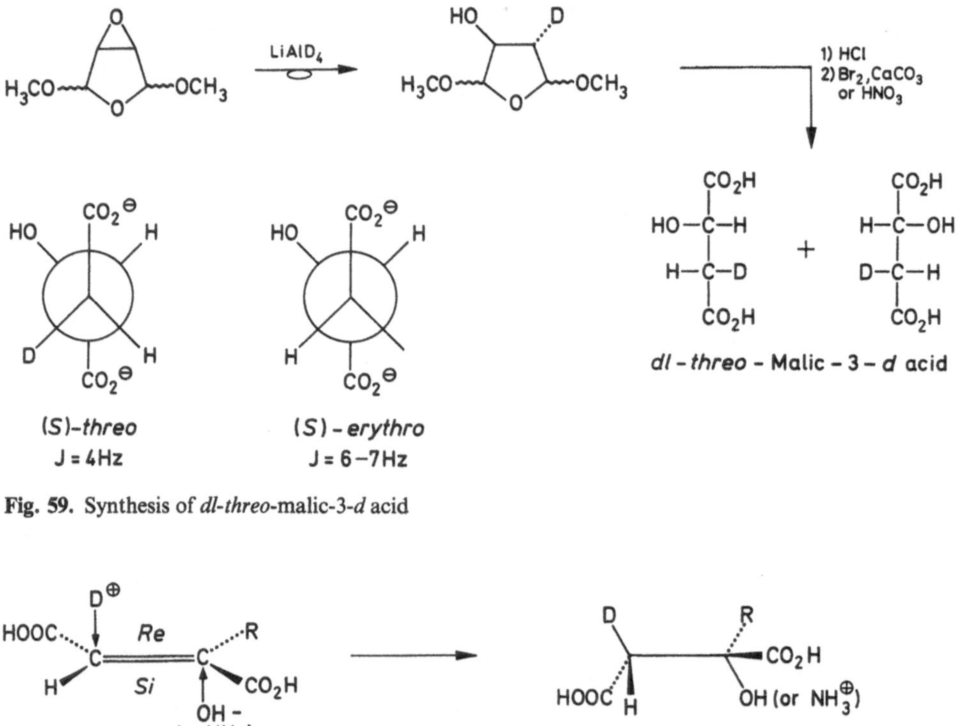

Fig. 59. Synthesis of *dl-threo*-malic-3-d acid

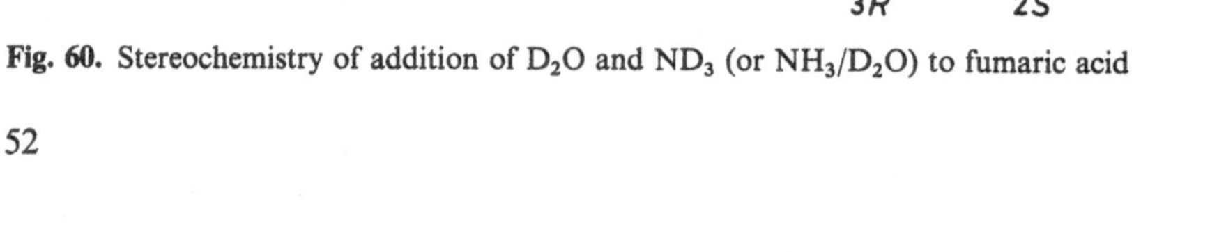

Fig. 60. Stereochemistry of addition of D$_2$O and ND$_3$ (or NH$_3$/D$_2$O) to fumaric acid

at least in principle, distinguishable by proton NMR spectroscopy. These hydrogen atoms will also be distinguishable by their reactivity in non-enzymatic as well as enzymatic, reactions. However, not only are enzymatic reactions likely to be more selective than other types of reactions as between diastereotopic groups (because of the already mentioned multiple interaction of the substrate with the enzyme and its active site) but, in addition, since the enzyme is chiral, it also "sees" the enantiotopic ligands H_A and H_C (or H_B and H_D) in a diastereotopic environment and thus can distinguish between them. It is therefore to be expected that the dehydration of citric acid mediated by aconitase will affect only one of the four hydrogens and indeed this has been found to be so [132]: when citric acid is equilibrated with isocitric acid *via* aconitic acid in the presence of aconitase (Fig. 61) but with D_2O instead of H_2O, only one of the four methylene positions acquires deuterium, hence only one of the four hydrogens can be eliminated in the dehydration step to aconitic acid and replaced in the reverse reaction.

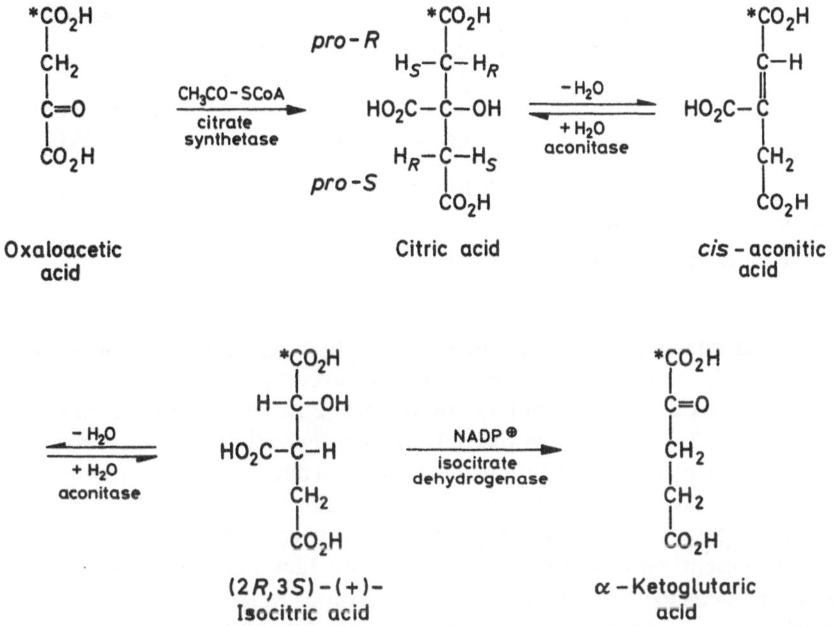

Fig. 61. Part of tricarboxylic acid cycle

The problem as to which hydrogen this is can be factorized into two parts:
1) Does the hydrogen come from the *pro-R* or the *pro-S* CH_2CO_2H branch of the citric acid? [This problem is, in itself, interesting since it is known from carbon labeling experiments discussed already in Section 2 (cf. Fig. 6) that the aconitase-active branch is also the one derived from the oxaloacetic acid, cf. the asterisks — for labeled carbons — in Fig. 61.]
2) Within the branch specified is the hydrogen abstracted H_R or H_S?

The first problem was solved [133] through synthesis (followed by enzymatic degradation) of tritiated citric acid stereospecifically labeled in the *pro-R* branch, as indicated in Fig. 62. The starting material for this synthesis is the naturally occurring 5-dehydroshikimic acid of known configuration (Fig. 62). Enzymatic hydration of the double bond in this acid with tritiated water gives the corresponding dihydro compound whose configuration, at C(1), is R.[32]

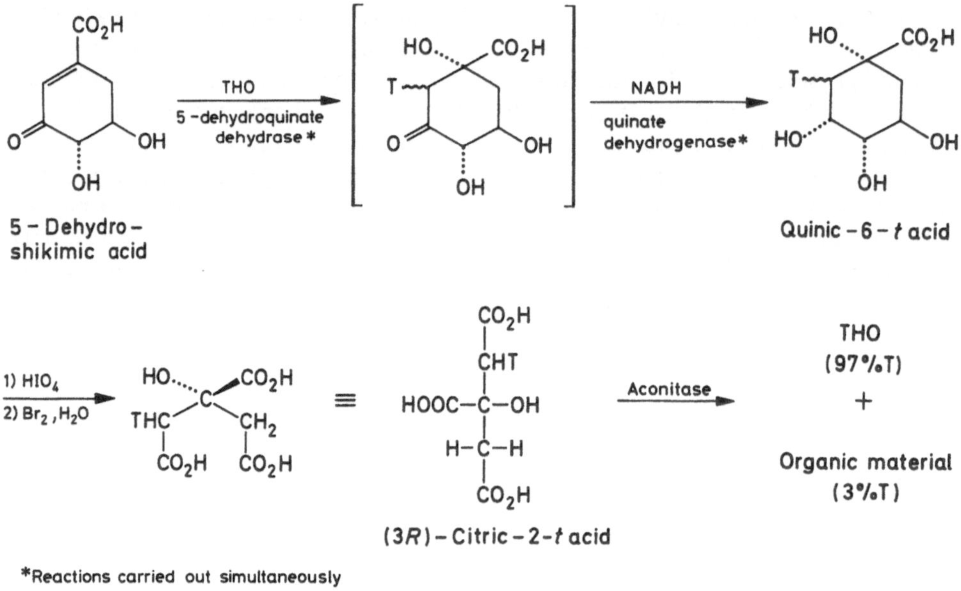

Fig. 62. Stereospecific synthesis of (3R)-citric-2-t acid

The hydrated material, without isolation, is reduced by NADH in the presence of a second enzyme to (1R)-quinic-6-t acid which is then cleaved by periodic acid to a dial-dehyde subsequently oxidized by bromine water to citric-2-t acid. While the configuration of this acid at C(2) is not known [because the configuration of its precursor at C(6) was not determined], its configuration at C(3) must, from the method of synthesis, be R. The label is therefore in the *pro-R* branch. Although the citric acid obtained is chiral through labeling, it will not, of course, display optical activity, since the tritium label is present only at the tracer level. The fact that the stereospecific labeling cannot be detected by ordinary chemical methods makes it no less real and when the labeled citric acid is treated with aconitase, the label indeed manifests itself in that the aconitic (and/or isocitric) acid produced is nearly free of tritium whereas the water eliminated carries nearly all the label. Since the label was in the *pro-R* branch (or better, the branch that would be *pro-R* in the unlabeled analog), it follows that the *pro-R* branch is the aconitase-active one. The same conclusion has been independently reached by a totally different method [134].

---

[32] By way of an exercise in nomenclature it should be noted that the configuration is R because the carbonyl-substituted segment of the ring has precedence over the hydroxyl-substituted segment. The configurational symbol is independent of the tritium labeling; it would be R in the corresponding dihydro analog as well, as required by the sequence rules [1]. However, in quinic-6-t acid, the configuration is R because the applicable sequence rule here demands that the labeled branch precede the unlabeled one.

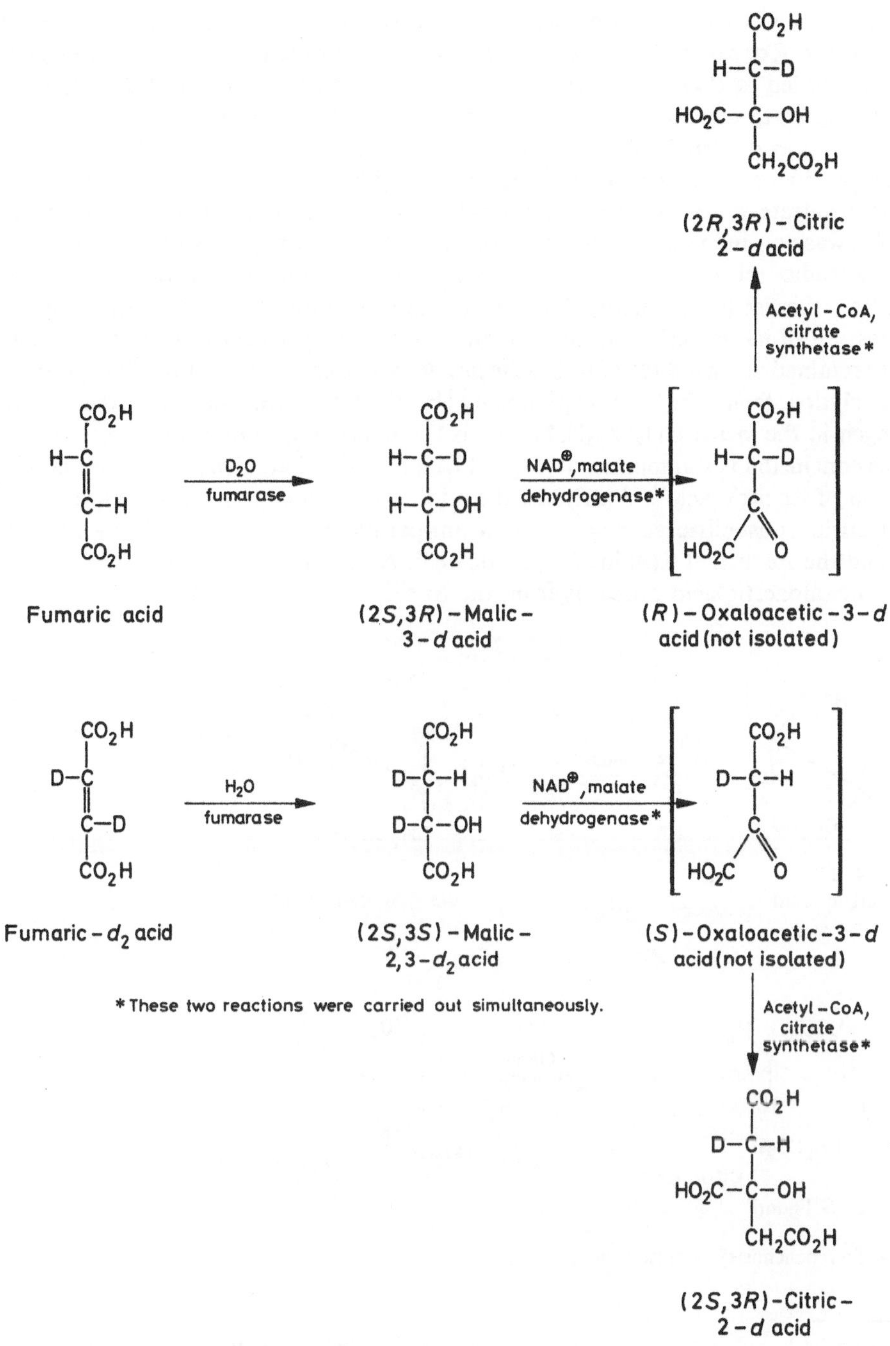

Fig. 63. Stereospecific labeling of citric acid at C(2)

We turn now to the question whether the hydrogen removed within the *pro-R* branch is *pro-R* or *pro-S*. An earlier performed sequence of reactions, shown in Fig. 63, had already led to that information [135]. (2S,3R)- and (2S,3S)-malic-3-d acids were synthesized by the method discussed earlier (Fig. 58) or an extension thereof. These acids were then oxidized enzymatically to oxaloacetic acid-3-d (3R or 3S, depending on the precursor), and the oxaloacetate was condensed *in situ* with acetyl-CoA by means of citrate synthetase to give citric-2-d acid. Each of the two citric-2-d acids (Fig. 63) was now incubated with aconitase. Since it was already known[14] from carbon labeling studies (cf. Fig. 6) that the oxaloacetic derived branch is the one affected by aconitase, success in this incubation was guaranteed. Indeed, the (2R,3R)-citric-2-d acid lost all of its deuterium in the dehydration to aconitic acid whereas the 2S,3R isomer retained at least 80% of it, i.e. the *pro-R* hydrogen at C(2) is lost. Thus, it may be concluded from the two experiments [133, 135] taken together that the *pro-R* hydrogen in the *pro-R* CH$_2$CO$_2$H branch is the one labilized by aconitase.[33]

This conclusion is summarized in Fig. 64 which shows that, given the known configuration of isocitric acid [111] (2R,3S), the addition of water to aconitic acid to give either citric or isocitric acid proceeds in antiperiplanar fashion to the *Re* face at C(2) and the *Re* face at C(3) in *cis*-aconitic acid. And, finally, the addition of acetyl-CoA to oxaloacetic acid proceeds from the *Si* side of the carbonyl function.[34]

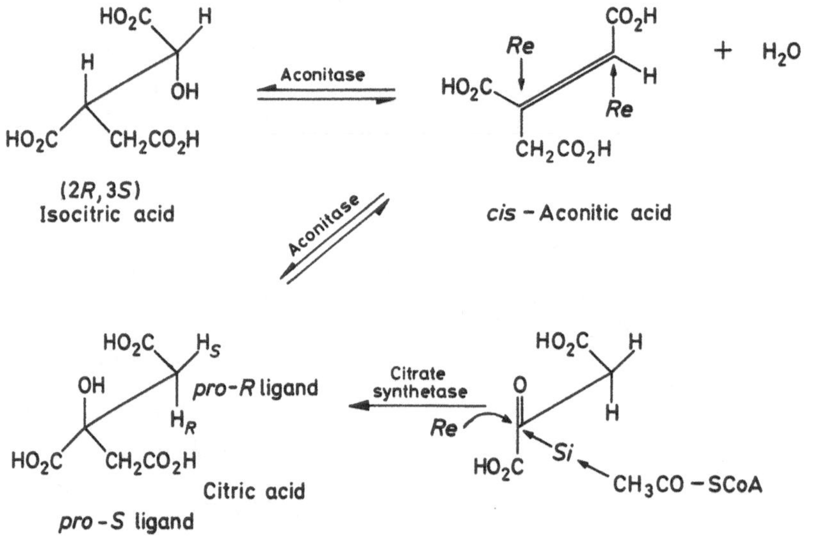

Fig. 64. Stereochemistry of citric acid cycle

---

[33] Working backward from the results, it may also be concluded that the configuration of the quinic-6-t acid (Fig. 62) at C(6) is R and that the enzymatic addition of THO to 5-dehydroshikimic acid is syn.

[34] This is not true for all citrate synthetases; enzymes from some sources lead to addition to the *Re* face [136]. This finding that enzymes from different sources promote one and the same reaction in stereochemically distinct fashion is by no means unique.

Many additional examples of the elucidation of prostereoisomerism in biochemical reactions could be given, for example the elegant elucidation by Cornforth and co-workers [111, 118, 137] of the biosynthesis of squalene, which was recognized by the Nobel prize in chemistry in 1975, or the recent studies of the enzymatic decarboxylation of tyrosine [138] and histidine [139] and of the condensation of homoserine with cysteine to give lanthionine [140], but the examples already provided should illustrate the principles and techniques involved in such studies.

All the examples given so far in this Section involve heterotopic hydrogen atoms and elucidation of stereochemical reaction course by use of deuterium or tritium. We shall conclude by providing two examples involving other heterotopic atoms, one concerned with carbon (^{12}C and ^{13}C) and one involving oxygen (^{16}O and ^{18}O).

The biosynthesis of the β-lactam antibiotic penicillin (Fig. 65), and also of cephalosporin, involves incorporation of L-valine and the question arises as to which of the two diastereotopic terminal methyl groups of the valine occupies which position in the penicillin. (In the case of cephalosporin, the question is as to which methyl group is incorporated into the six-membered ring and which becomes the methylene group of the carbinyl acetate.) The problem has been solved by two groups [65d, 141] by synthesis of specifically ^{13}C methyl labeled valine (cf. Fig. 42, and p. 35) which was then biosynthetically incorporated in the antibiotics. The position of the ^{13}C in the resulting antibiotic molecules was determined by ^{13}C NMR spectroscopy.

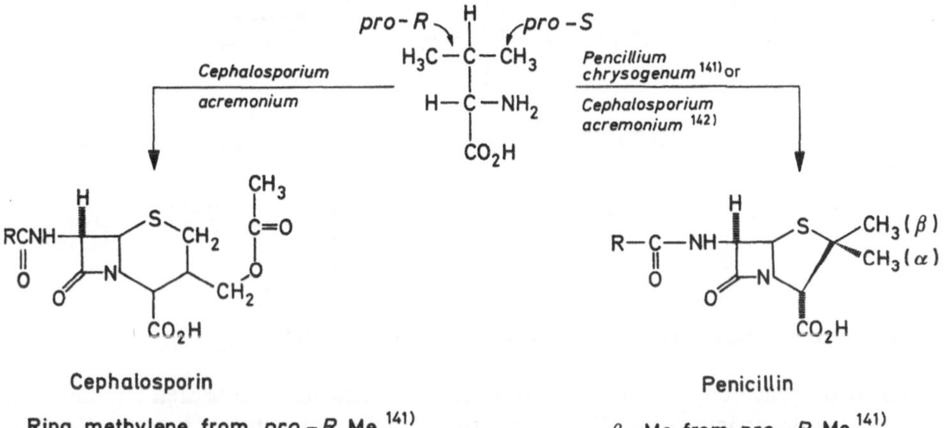

Ring methylene from *pro-R* Me [141]
Side-chain methylene from *pro-S* Me [65d]

β-Me from *pro-R* Me [141]
α-Me from *pro-S* Me [65d]

Fig. 65. Biosynthesis of penicillin[35] and cephalosporin from valine[36]

---

[35] The penicillin synthesized in Ref. 141 was penicillin V, R = $C_6H_5OCH_2$—. That synthesized in Ref. 65d was penicillin N, R = (*R*)—$HO_2CCH(NH_2)CH_2CH_2$—. The latter R-group was also present in both of the cephalosporin samples synthesized (cephalosporin C).

[36] The valine used in Ref. 65d was *pro-S* labeled, i.e. it was (2*S*,3*S*)-valine-4-^{13}C. The valine used in ref. 141 was racemic at C(2) but stereospecifically labeled at C(3): it was (2*RS*,3*R*)-valine-4-^{13}C. [In the original communications [65c, 141] the configuration was erroneously denoted as (2*RS*, 3*S*).] The presence of the 2*R* isomer is immaterial, since it is not biosynthetically incorporated.

The results are summarized in Fig. 65.[36] The position of labeling of the cephalosporin follows directly from the position, in the ^{13}C spectrum, of the peak enhanced by labeling when the precursor is stereospecifically methyl-^{13}C labeled valine. The position of labeling of the penicillin, α or β, was deduced on the basis of an earlier assignment [142] of the two signals due to the diastereotopic methyl groups in the spectrum of the unlabeled material.

An application of heterotopic oxygen atoms is elegantly illustrated by the elucidation of the stereochemical course of ring openings of CAMPS (the thio analog of cyclic AMP) with water to give AMPS, the thio analog of AMP [143]. The reaction sequence is shown in Fig. 66.

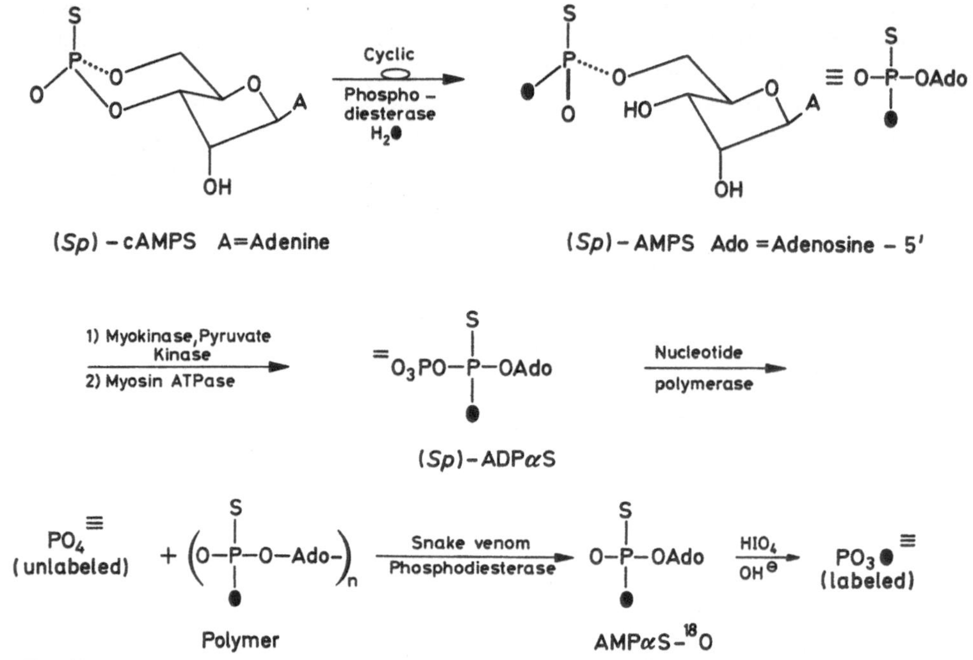

Fig. 66. Stereochemistry of enzymatic hydrolysis of cAMPS to AMPS

The two diastereomers of cAMPS were stereospecifically synthesized [145] and their configurations assigned by ^{31}P NMR, the S_p (the subscript indicating that the configuration symbol refers to phosphorus[37]) isomer with equatorial oxygen having the more upfield ^{31}P shift [144]. The cAMPS was ring-opened with $H_2^{18}O$ in the presence of phosphodiesterase from beef heart. It may be seen from Fig. 66 that, if this ring-opening proceeds with inversion (as indicated in the figure), the product

---

[37] The following points should be made here: a) In the sequence rules, isotopic differences are to be considered only after all other material differences are exhausted, thus $S > \ddot{O}P > \ddot{O}C > ^{18}\ddot{O}$: but $^{18}\ddot{O}^{\ominus} > ^{16}\ddot{O}^{\ominus}$. b) Resonating P=O double bonds and negative charges (and sometimes even labile protons) are generally disregarded; thus in $>P{<}^O_{O^-}$ $\left(\text{and possibly even in } >P{<}^O_{OH}\right)$ the two oxygens are on a par and if one is labeled, it is given precedence in the sequence rule.

AMPS-^{18}O is the S or P_S isomer — or putting it differently, in an unlabeled sample the *pro-S* oxygen is derived from the water and the *pro-R* oxygen from the cAMPS. (Contrariwise, were the hydrolysis to proceed with retention, the R or P_R isomer of AMPS would be obtained.) To analyze the AMPS, it was first diphosphorylated to ATPαS by means of phosphoenolpyruvate in the presence of myokinase and pyruvate kinase. This reaction is known [145] to involve stereoselective phosphorylation of the *pro-R* oxygen. If the reaction course is as shown in Fig. 66, this is the unlabeled oxygen, (whereas it would be the labeled one if cAMPS hydrolysis had involved retention). The two species can actually be distinguished because the ^{18}O isotope produces a chemical shift differential of the adjacent ^{31}P nucleus [145a, 146]. In the ATPαS actually obtained, since the bridging oxygen is ^{16}O, only the phosphorus directly attached to the adenosine (P_α) should display the shift, whereas in the opposite situation, where ^{18}O would have been bridging, both the first and the second phosphorus nuclei (P_α and P_β) should have been shifted. In actual fact, however, a different method of analysis was employed [143]. The ATPαS was cleaved to ADPαS with myosin ATPase and the ADPαS was then polymerized by means of nucleotide polymerase. In this polymerization the outer (β) phosphate of ADPαS is extruded with its bridging oxygen and the inner phosphate is linked to the 3'-hydroxyl of another ADPαS molecule, thus forming a polymer with alternating adenosine and phosphodiester units, the 5'-position of one adenosine being linked by the phosphate to the 3'-position of another. It can be seen from Fig. 66 that, if the unlabeled oxygen forms the bridge in the diphosphate (as shown), the extruded phosphate is unlabeled and the phosphate linkage in the polymer is ^{18}O-labeled. (If the bridging oxygen were the labeled one, then the extruded phosphate would be labeled and that remaining in the polymer unlabeled.) In the event, the free phosphate formed in the polymerization was methylated and the resulting trimethyl phosphate analyzed mass spectrometrically: it had virtually no excess ^{18}O. By way of control the polymer was degraded by treatment with snake venom phosphodiesterase followed by cleavage with sodium periodate and alkali. This freed the bound (linking) phosphate, which was, in turn, converted into trimethyl phosphate and analyzed by mass spectrometry. It was found to contain a considerable amount of excess ^{18}O. Reasoning backward, the following conclusions were reached [143]: The phosphate unit in the polymer, i.e. the unlinked oxygen in the α-phosphate unit of ADPαS, is labeled; the β-phosphate unit is unlabeled. Since phosphorylation of AMPS occurs at the *pro-R* oxygen [144] the *pro-R* site is unlabeled and the *pro-S* site is therefore labeled; in other words, the AMPS formed in the hydrolysis of (S_p)-cAMPS is the S-isomer, as shown in Fig. 66. Contemplation of the first step in that figure thus indicates that the hydrolysis of cAMPS proceeds with inversion of configuration at phosphorus.

The reverse reaction, cyclization of ATPαS to cAMPS, also proceeds with inversion of configuration, as shown by Gerlt and coworkers [147].

5.3 Chiral Methyl and Phosphate Groups

5.3.1 Chiral Methyl Groups

The earlier-described (p. 53) elucidation of the stereochemistry (or prostereochemistry) of the acetyl-CoA — oxaloacetate reaction is incomplete in one respect: it does not disclose whether reaction of the methyl group of CH_3CO-CoA proceeds with retention or inversion. Posing the question in this way may be momentarily puzzling, but it will be recognized that if the condensation, instead of involving acetate, involved the metabolic poison fluoroacetate, $FCH_2CO_2^-$, the methylene carbon in this compound would be prochiral and one could ascertain whether the *pro-R* or *pro-S* hydrogen were involved in the carbanion formation by working with chiral $FCHTCO_2^-$. Also, by determining the configuration of the fluorocitrate formed, one could find out whether the condensation proceeds with retention or inversion [148]. In principle, the same considerations should apply to deuterioacetate, $DCH_2CO_2^-$ in which the two hydrogen atoms have become enantiotopic. Stereospecific labeling with tritium would then require chiral $CHDTCO_2H$. This section will deal with the synthesis of this material and its use in enzymatic reactions [149]. The latter subject is complicated by the fact that, unlike in the fluoroacetate case, hydrogen abstraction to give the intermediate carbanion may occur with any one of the three isotopes.[38]

The synthesis of $CHDTCO_2H$ is, in principle, straightforward and was first accomplished in two laboratories in 1969 [150]. A modified version of the Cornforth synthesis [151] is shown in Fig. 67. Most of the steps are self-evident. The doubly labeled phenyl methyl carbinol is, of course, a *dl* pair in the first instance: the absolute stereochemistry at the CHDTX chiral center is *RS* but the relative stereochemistry of the two chiral centers is defined as $1RS,2RS$, i.e. only one of the two possible diastereomers is obtained. The resolution at the methyl group occurs automatically when the carbinol is resolved by classical methods.

The original method described by Arigoni's group [150b] involved generation of the chiral center by enzymatic means, but later an elegant, purely chemical method for the synthesis of chiral $CHDTCO_2H$ with high specific radioactivity was described by the Zurich investigators [152]. Other methods for the synthesis of chiral CHDTX compounds have since been developed [149, 153].

The configuration, enantiomeric purity or even the chiral nature of the (*S*)-$CHDTCO_2H$ cannot be established by classical methods (although the fact that it is chiral, and its configuration, follow from the method of synthesis). Quite apart from the question as to whether a chiral center of type CHDTX would display measurable rotation, the $CHDTCO_2H$ is diluted at least thousandfold by carrier CH_2DCO_2H since the tritium is present at the tracer level. Thus even if the pure material had a

---

[38] Centers of the type CH_3CO_2H or $ROPO_3^=$ have been called [149] "pro-prochiral". If this term denotes a merely formal relationship — namely that replacement of a *a* at a "pro-prochiral" center CaaaX by Y gives a prochiral center CaaXY — it is probably unobjectionable. However, unlike, at a prochiral center where the heterotopic ligands are, in principle, distinguishable by enzymatic or spectroscopic means, the (homotopic) ligands at a pro-prochiral center are operationally indistinguishable except possibly at the slow rotation limit where the three homomorphic ligands *a* in (i) will become heterotopic. This has been observed in certain cases where *a* = CH_3: Suzuki, M., Ōki, M., Nakanishi, H., Bull. Chem. Soc. Japan, *46*, 2858 (1973), or *a* = F, see Fig. 48.

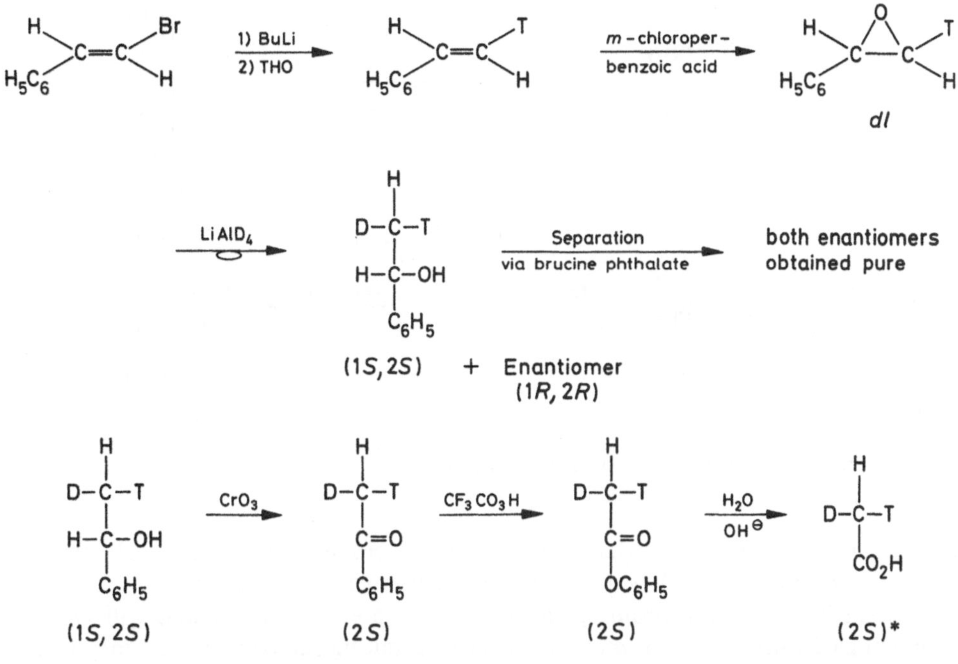

* The 2*R* enantiomer is obtained analogously from the enantiomeric (1*R*,2*R*) precursor

Fig. 67. Synthesis of chiral $CHDTCO_2H$

small rotation, it would not be detectable at the existing dilution.[39] Recognition of configuration thus presents a serious problem whenever the chiral acetic acid is obtained as a reaction product. The problem was solved [150)] by condensing the acetic acid (as Co-A ester) with glyoxylate to yield malate and then diagnosing whether the malate formed had the tritium in the *pro-R* or *pro-S* position, by the method already outlined in Fig. 58. The reaction sequence utilized is shown in Fig. 68.

Because of the rapid rotation of the methyl group in $CHDTCO_2H$, the question as to whether H, D or T is abstracted in the formation of the carbanion or carbanionoid intermediate in the aldol condensation to give malate is not determined stereo-

[39] For the use of tritium at a high level of incorporation and detection of its stereochemical placement in a CHTXY* group by tritium NMR, see ref. 154. This method does not seem applicable to CHDTX* (the stars refer to chirality in X or Y), however, since the steric isotope effect of D would presumably be insufficient to induce enough of a population difference (cf. Sect. 4.3) among the rotameric conformations of —CHDT to generate a palpable excess of tritium in one or other of the two diastereotopic placements i or ii. Nor would it appear likely that the intrinsic shift difference (cf. Sect. 4.3) induced by the exchange of protium and deuterium would be sufficiently large to show a difference between the tritium nuclei in the two diastereotopic loci, though this is perhaps somewhat less certain.

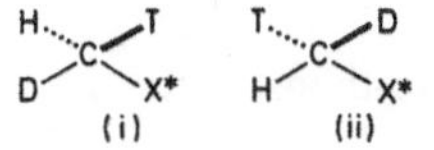

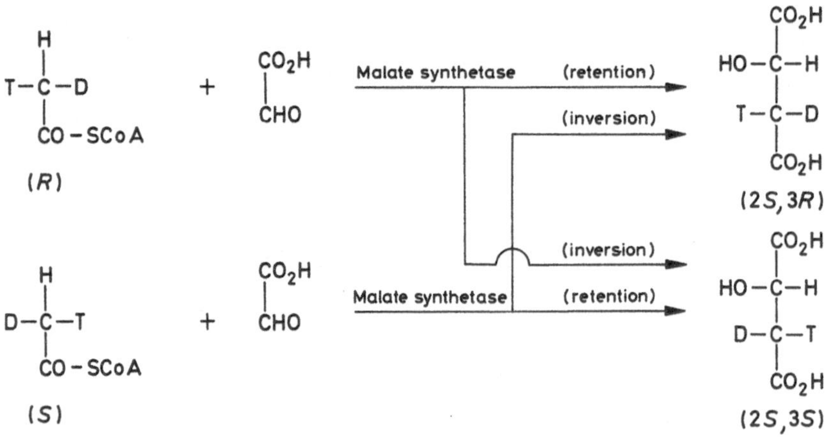

Fig. 68. Stereochemistry of malate synthetase reaction

chemically but by an isotope effect. If the isotope effect is normal, ease of abstraction will be H > D > T. Fortunately, tritium abstraction may be disregarded, for when it occurs, the resulting malate will be non-radioactive and will become commingled with the carrier material as far as counting radioactivity is concerned.[40]

If hydrogen is abstracted preferentially over deuterium, then retention of configuration in the condensation means that (R)-acetyl-CoA gives (3R)-malate-3d, t and (S)-acetyl-CoA (3S)-malate-3d, t (Fig. 68, top and bottom lines). But if inversion is the course, the (R)-acetyl-CoA gives (3S)-malate-3d, t and (S)-acetyl-CoA (3R)-malate-3d, t (Fig. 68, middle lines). These conclusions are, as already indicated, predicated on a normal isotope effect, i.e. preferred abstraction of H over D from the acetyl-CoA.

To diagnose the situation, the malates formed were treated with fumarase. The one formed from (R)-acetyl-CoA became equilibrated with tritiated fumaric acid and retained most of its tritium whereas that from the (S)-acetyl-CoA lost most of its tritium in the reversible dehydration yielding unlabeled fumaric acid (and, then, unlabeled malic acid). Reference to Fig. 58 shows that the former malate was thus 3S and the latter 3R, i.e. the condensation proceeds with *inversion* of configuration.

The extent of retention of tritium is 79 ± 2% in the former case and 21 ± 2% in the latter [155]; complete retention (or complete loss) of tritium is, of course, not to be expected, since the isotope effect for extrusion of deuterium rather than protium in the condensation shown in Fig. 68 is not infinite. The percent tritium retention in the process of dehydration and water exchange (fig. 58) is now generally called the "F-value" [149]; thus enantiomerically pure $CHDTCO_2H$ gives an F-value of either 79 (R configuration) or 21 (S configuration); values below 79 or above 21, i.e. closer to 50, indicate that the acid is not enantiomerically pure. The F-value determination

---

[40] However, the observed [150a] fact that the malate contains more than two-thirds of the tritium of the acetate is *prima facie* evidence of a primary isotope effect.

thus takes the place of the classical determination of enantiomeric purity (e.g. by measurement of rotation, or by NMR spectroscopy using chiral shift reagents) used for conventional chiral compounds of type CHabc and CHDab. (Other methods for determining the configuration and enantiomeric purity of $CHDTCO_2H$ have been described [149] but are less convenient and not often used.)

The ability to analyze $CHDTCO_2H$ stereochemically was put to use in the elucidation [156, 157] of the stereochemistry of the citric acid condensation to which we have already alluded. Before presenting one of the determinations of the stereochemistry of the citrate synthetase reaction, however, we must develop that of the reverse process, the citrate lyase reaction [156]. The essential features are shown in Fig. 69, bottom. The key to this determination is the synthesis of citric acid stereo-

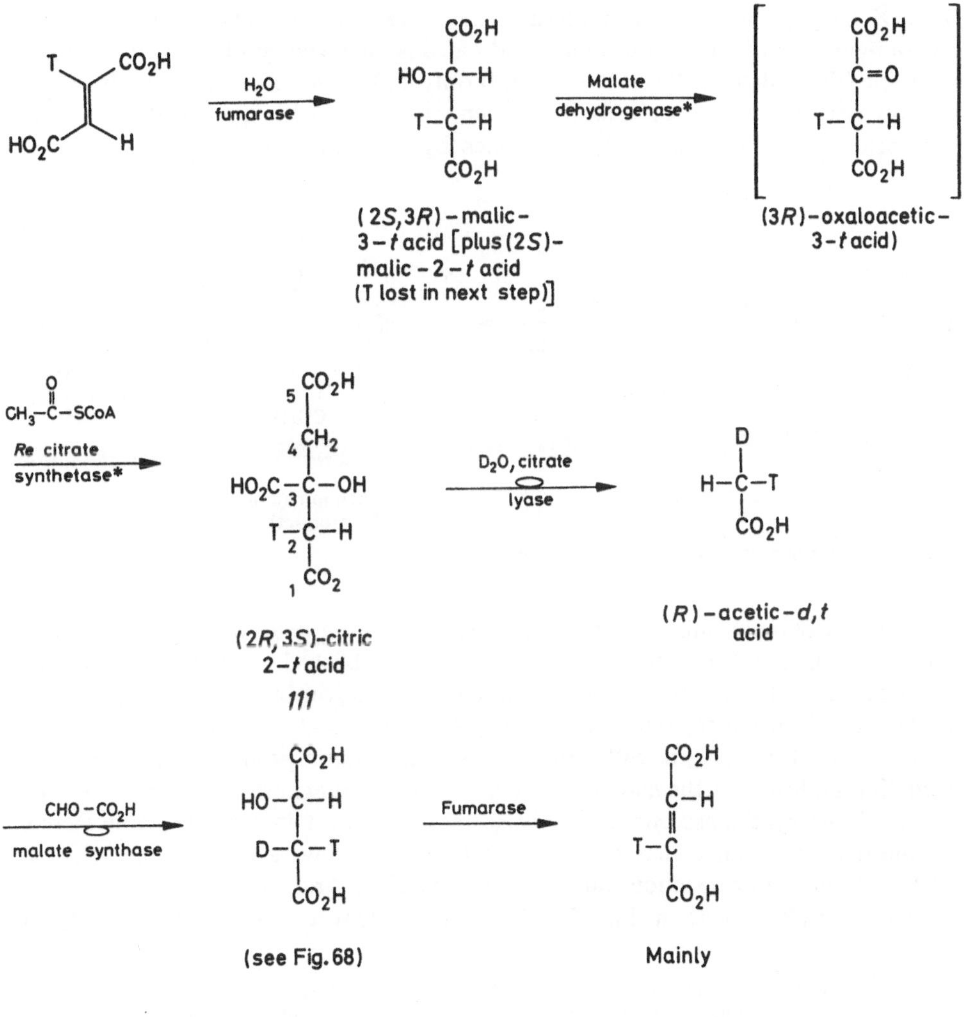

*steps performed simultaneously

Fig. 69. Stereochemistry of citrate lyase reaction

specifically labeled at the methylene group in the *pro-S* branch [the carbon-1,2 branch of the (2*R*,3*S*)-citric-2-*t* acid (*111*) Fig. 69]. At first sight this seems to be difficult to accomplish, at least enzymatically, because that branch is the acetyl-CoA derived one (cf. Fig. 63). However, it was possible to get around this difficulty by using *Re*-citrate synthetase (rather than the usual *Si*-citrate synthetase, see footnote 34 on p. 56) to prepare the citric acid from ordinary acetyl-CoA and stereospecifically labeled oxaloacetic acid, as shown in the top part of Fig. 69. With this accomplished, the stereospecifically labeled (2*R*,3*S*)-citric-2-*t* acid was subjected to the citrate lyase reaction and the acetic-*d*,*t* acid formed was diagnosed to be *R* by the already explained malate synthetase/fumarase sequence. It follows that the citrate lyase reaction proceeds with *inversion* of configuration, as shown in Fig. 69. It should be noted that this result is *independent* of any assumptions on isotope effects, for while such effects are involved in the transformation XCHDT→XCYDT (rather than XCHYT), they are *not* involved in the reverse process. (The diagnosis of the *R*-configuration of any sample of acetic-*d*,*t* acid is independent of the stereochemical course of the malate synthetase reaction. It only makes use of the fact that the course of this reaction is the same as it is when one starts with known *R*-acid whose stereochemistry is unequivocally determined by its synthesis.)

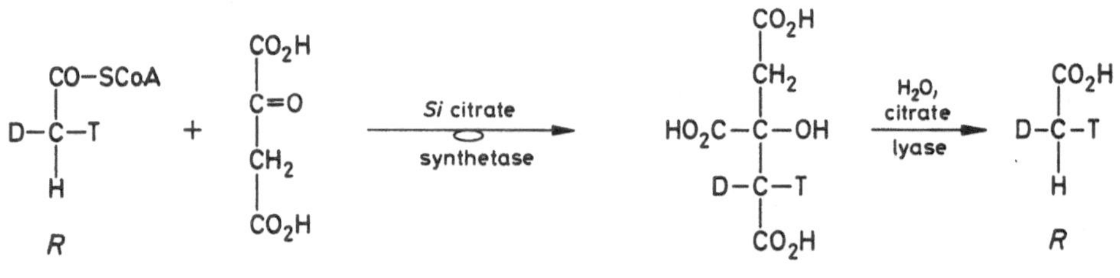

Fig. 70. Stereochemistry of *Si* citrate synthetase reaction

With the stereochemistry of the citrate lyase reaction determined, that of the *Si* citrate synthetase (the common enzyme) was established as shown in Fig. 70. Condensation of (*R*)-acetic-*d*,*t* acid (configuration known by synthesis) with oxalo-acetate gives what turns out to be mainly (2*S*,3*R*)-citric-2-*d*,2-*t* acid (*112*).[41] When this acid is then cleaved with citrate lyase, the major product is (*R*)-acetic-*d*,*t* acid, as established by the malate synthetase/fumarase diagnosis. It follows that both the *Si*-citrate synthetase and citrate lyase reactions must involve the same stereochemical course. Since that of the lyase reaction is inversion *(vide supra)*, that of the *Si* synthetase reaction must be inversion also. And since the overall stereochemical result shown in Fig. 70 is not dependent on the magnitude of the

---

[41] Because the H/D isotope effect is not infinite, there will also be some (2*R*,3*R*)-citric-2-*t* acid in which the deuterium was abstracted from the acetyl-CoA. This material will give CH_2TCO_2H in the citrate lyase reaction which, being achiral, behaves like racemic $CHDTCO_2H$ in the subsequent steps.

isotope effect,[42] neither is the demonstration of inversion in the Si citrate synthetase reaction.

Because of the existence of a recent, thorough review [149] we shall not summarize the by now quite extensive literature on chiral methyl groups in detail. Suffice it to point out [149] that there are three ways in which chiral methyl groups have been used in stereochemical studies, mostly in bioorganic chemistry. (There is no reason why the CHDTX group could not also be used in classical mechanistic organic studies, except possibly that the practitioners of such studies may be unfamiliar with the enzymatic methodology required in the diagnosis of configuration.) The first instance, exemplified by the malate synthetase (Fig. 68) and citrate synthetase (Fig. 70) reactions, involves a transformation of the type CHDTX→CDTXY. It is in this transformation that intervention of an isotope effect is required for observation of the stereochemistry, as explained earlier (but see below). Ordinarily one assumes that the isotope effect is normal, i.e. that H is replaced or abstracted faster than D. If this assumption is wrong, i.e. if there is an inverse isotope effect, then the answer obtained (retention or inversion of configuration) is the opposite of the correct one. Fortunately such instances are rare and can be guarded against in various ways. One is to check the isotope effect independently [155b]. A second approach [158] is applicable when the hydrogen isotopes in the CDTXY (or CHTXY) product are in an exchangeable position. Let us take the case depicted in Fig. 68, starting with (R)-CHDTCO-SCoA as a (hypothetical) example and let us assume that the reaction goes with inversion of configuration. Then, if a normal isotope effect is involved, the predominant product will be the $(2S,3S)$-3-d,t isomer (113), but if an inverse isotope effect were involved, the predominant product would be $(2S,3R)$-3-h,t (114) as shown in Fig. 71. Were the reaction to go with retention, the opposite result would be expected (Fig. 71, 115, 116). As already mentioned, the configuration at C(3) is determined in that the $3R$ isomer (114 or 115) loses tritium with fumarase and the $3S$ isomer (113 or 116) retains it. The test for the isotope effect [158] involves a partial exchange of the enolizable T, D or H at C(3) by H_2O and base. Material which loses tritium becomes unlabeled and is lost to further observation. Of the material which retains tritium, that which contains a CHT group will be racemized by exchange of H; that which contains a CDT group will also be racemized, but more slowly (relative to tritium loss) because base catalyzed enolization is known to have a sizeable *normal* primary isotope effect. Thus, in the case where the reaction course was inversion (fig. 71, left half), the $3R$ isomer (114) will be racemized faster than the $3S$ (113) and the relative amount of tritium retention upon subsequent fumarase treatment will initially increase. This result is independent of the isotope effect (direct or inverse) operative during the condensation step (fig. 68). But if the original reaction involved retention (fig. 71, right half), the $3S$ isomer (116) will be racemized faster than the $3R$ (115) and the relative amount of tritium loss upon fumarase mediated exchange will increase following the base/H_2O treatment, again regardless of the initial isotope effect in the glyoxylate — CHDTCOSCoA condensation. Thus the effect

---

[42] The lower the isotope effect, the less the preservation of optical purity in the overall reaction. However, even if there were an inverse isotope effect, the stereochemical outcome would not be altered; i.e. one would not obtain any excess of (S)-acetic-d,t acid in the end, rather the (R)-acid would be isotopically diluted by much CH_2TCO_2H.

```
   CO₂H              CO₂H              CO₂H              CO₂H
    |                 |                 |                 |
HO—C—H            HO—C—H            HO—C—H            HO—C—H
    |                 |                 |                 |
 D—C—T             T—C—H             T—C—D             H—C—T
    |                 |                 |                 |
   CO₂H              CO₂H              CO₂H              CO₂H

  (2S,3S)           (2S,3R)           (2S,3R)           (2S,3S)
   113               114               115               116
  normal            inverse           normal            inverse
        isotope effect                      isotope effect
          Inversion                           Retention
```

Fig. 71. Predominant labeled malates

of the H/D exchange upon the shift in outcome of the final fumarase mediated exchange is in the same direction, whether the isotope effect is normal or inverse, and depends only on the stereochemistry of the malate synthetase condensation.[43] In fact, what the exchange experiment tells one is which of the two stereoisomers at C-3 has H next to T and which D; it is interesting that, even if the acetate-glyoxylate condensation (Fig. 68) involved no isotope effect at all ($k_H/k_D = 1$), the exchange test could reveal the stereochemistry of the condensation [158]. A third way of evading the isotope problem is to look at both the forward and reverse reactions; the stereochemistry of the latter (see below) can be studied without reference to an isotope effect and that of the former then follows. The citrate lyase — citrate synthetase case (Fig. 70) illustrates this approach. A fourth approach involves recognition of the stereochemistry of the product by tritium NMR [154]. In that case, not only will the chemical shift of a CXYHT group differ from that of the stereochemically analogous CXYDT group (because of an isotope shift), but also the former, but not the latter, will show a tritium doublet in the proton-coupled spectrum. The two case are thus readily distinguished and the relative intensity of the signals will reveal if the isotope effect is normal or inverse. An application is in the biosynthesis of cycloartenol (Fig. 72) [154]; the cyclization of methyl to methylene proceeds with retention of configuration. [This result has been independently confirmed by Blättler and Arigoni using different methodology and an enzyme preparation from *Zea Mays*. [159)

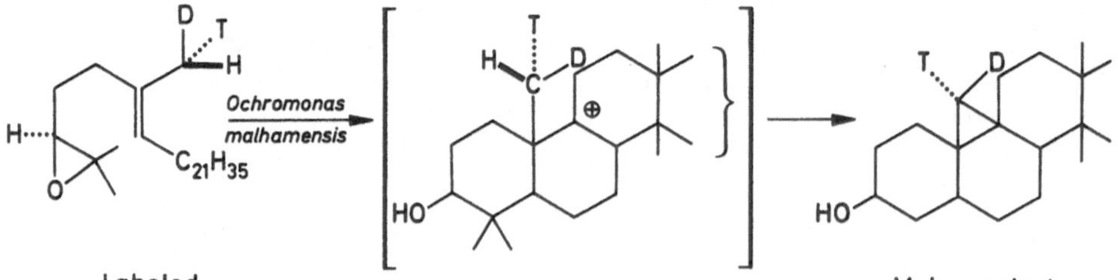

Fig. 72. Stereochemistry of cycloartenol biosynthesis

---

[43] The actual example [158)] related to fatty acid biosynthesis; the case of malate synthetase here discussed is hypothetical.

A number of other transformations of the CHDTX→CDTXY type have been tabulated [149]; most of them involve retention of configuration, with the notable exception of Claisen condensations *(videsupra)* which all go with inversion. A recent example [160] is the hydroxylation of $CH_3(CH_2)_6CHDT$ to $CH_3(CH_2)_6CDTOH$ by *Pseudomonas oleovorans*; this reaction proceeds with retention of configuration.

A second type of reaction whose stereochemistry was elucidated by the use of chiral methyl groups is of the CDTXY→CHDTX type, exemplified by the citrate lyase reaction (Fig. 69). The stereochemical outcome of a number of these reactions has also been tabulated [149].

A third type of reactions to be studied by chiral methyl groups is of the methyl transfer type: CHDTX→CHDTY. As one might expect, these reactions often involve inversion but not invariable so [149]. An example is the catechol-O-methyltransferase (COMT) reaction shown in Fig. 73. [161] The important methyl transfer reagent S-adenosylmethionine with a stereochemically labeled methyl group (*117*) is synthesized [161] from chiral acetate via a Schmidt reaction followed by tosylation to N-tosylmethylamine CHDTNHTs which is converted to the ditosylate in which the $N(Ts)_2$ group becomes a leaving group. Reaction with the S-sodium salt of homocysteine gives methionine(CHDT) which is converted to the adenosyl derivative by means of ATP. Analysis of the product involves oxidation of the benzene ring with ceric ammonium nitrate to CHDTOH, followed by sulfonylation, treatment with cyanide and hydrolysis to $CHDTCO_2H$ which is analyzed as described earlier. Both the preparation and the analysis involve one inversion step; this is, of course, taken into account in the consideration of the stereochemistry of the COMT catalyzed reaction which involves inversion. The methylation of the polygalacturonic acid carboxyl groups of pectin catalyzed by an enzyme preparation from mung bean shoots has been studied similarly [162].

5.3.2 Chiral Phosphate Groups

One other element which is conveniently available in three isotopic forms is oxygen. These isotopes have been used[44] to synthesize [163,164] chiral phosphates of the type

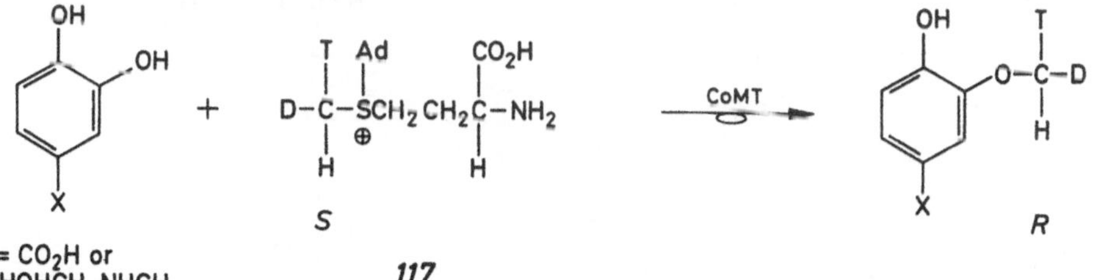

X = CO_2H or
–$CHOHCH_2NHCH_3$
(epinephrine)

117

Ad = Adenosyl

Fig. 73. Stereochemistry of catechol-O-methyltransferase mediated reaction

---

[44] Both Knowles [163] and Lowe [164] in 1978 synthesized phosphate monoesters (of 2-O-Benzyl-1,2-propanediol and methanol respectively) whose configurations could be inferred from the method of synthesis (but see below and [168]). Knowles [163] checked the configuration of his product by cyclization followed by an ingenious mass spectral analysis. Lowe [164] had no check of configuration but reported [164] that the chiral methyl phosphate, rather surprisingly, displayed measurable circular dichroism.

RO—P^{16}O^{17}O^{18}O$^=$ conveniently written as ROPOⒼ⚫$^{2-}$. There is one obvious disadvantage and one advantage of ROPOⒼ⚫$^{2-}$ over CXHDT: the disadvantage is that isotope effects would be very small and probably not practically useful as among the oxygen (as distinct from the hydrogen) isotopes; the advantage is that the P—O bonds are much easier to transform chemically in stereochemically defined ways than the C—H bonds and that ^{17}O (50% enriched) is much easier to use in high concentration than the radioactive T. (^{18}O, like D, is available in over 99% purity.)

Triply labeled phosphate has been employed to elucidate the steric course of a number of phosphoryl transfer reactions; the topic has been reviewed [165, 166], and we shall here present only one example, concerned with the stereochemistry of cyclization of ADP to cAMP [167, 168] and the reverse reaction [169, 170].

The pathway shown in Fig. 74 represents several syntheses of chirally P-labeled phosphate esters with different R's, including AMP, carried out by Lowe and his group [165]. Assignment of configuration of the P-labeled AMP depends on the proper assignment of configuration of the cyclic phosphate precursor 120, R=CH$_3$ [168] which was originally made incorrectly [163, 167] but has since been corrected [165, 168, 169]. Chiral benzoin (118) was prepared from resolved mandelic acid and was ^{18}O labeled by conversion to the ethylene ketal followed by cleavage with H$_2$^{18}O and acid. Reduction to chirally labeled hydrobenzoin (119) was followed by conversion to a cyclic, ^{17}O labeled phosphate of established [168] relative position of the =O and OR groups; the absolute configuration at phosphorus is thus S, as shown in Fig. 74, both before (120) and after (121) hydrogenolysis.

When R in 121 is 5'-adenosyl,[45] the product is P-chirally labeled AMP, adenosine-5'[(S)-^{16}O^{17}O^{18}O] phosphate (122). Its cyclization to chirally ^{16}O, ^{18}O

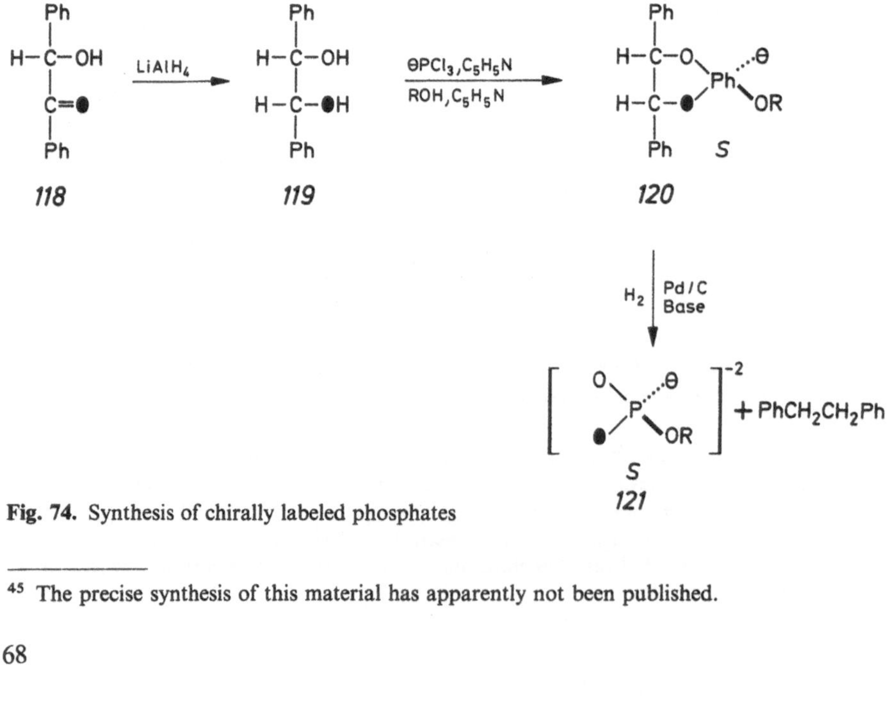

Fig. 74. Synthesis of chirally labeled phosphates

---

[45] The precise synthesis of this material has apparently not been published.

labeled cAMP is shown in Fig. 75. The NMR analysis of this material is based on the discovery of the very useful effects of ^{17}O (quenching) [171] and ^{18}O (isotope shift) [146,147,171b,172] on the shift of adjacent ^{31}P nuclei. Of the three products which can be formed by displacement of any one of the three oxygen atoms on phosphorus (the isotope effect is negligible in this case), one will contain ^{16}O and ^{17}O, one ^{17}O and ^{18}O and the third one ^{16}O and ^{18}O.

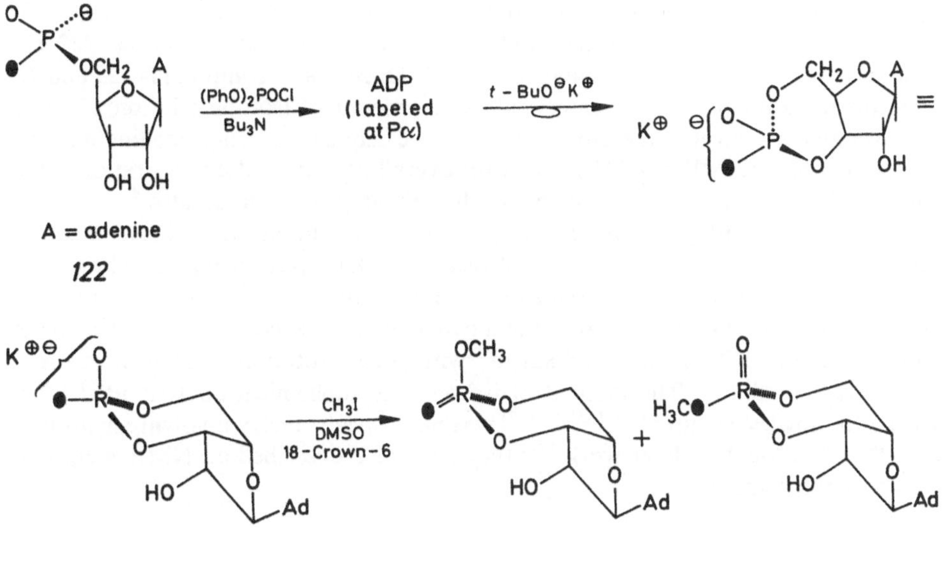

Fig. 75. Cyclization of stereospecifically P—O labeled AMP to cAMP

Any species which contains the P—^{17}O (P—Θ) moiety will have its ^{31}P resonance quenched by the large quadrupole moment of the adjacent ^{17}O nucleus [171]. The only cAMP species shown in Fig. 75 is therefore the one formed by displacement of ^{17}O, since other species will retain the P—Θ moiety and will not be seen in the ^{31}P NMR analysis used [167]. The course depicted in Fig. 75 is the actual one of inversion at phosphorus: if retention occurred, the ^{16}O and ^{18}O nuclei would be interchanged. Thus, in the final methylation product, the ^{18}O is necessarily equatorial, either as ●=P (*123*) or as CH$_3$●—P (*124*), depending on whether methylation occurs equatorially or axially. Indeed two families of ^{31}P signals are seen for the two diastereomeric species. Since an ^{18}O nucleus induces an isotope shift at an adjacent ^{31}P [172] and since, moreover, this shift is different for a doubly bonded ● than for a singly bonded one, species *123* will show the isotope shift typical for ●=P whereas *124* will show the shift for —●—P.[46] (If the reaction had gone with retention, the labeled (^{18}O) oxygen would be axial in which case species *124* would have displayed

---

[46] Because ^{17}O can be obtained in up to 50% purity only, there will always be some unlabeled species present which can serve as calibration or reference points for the isotope shifts.

Ernest L. Eliel

the ● = P isotope shift and *123* the —●—P single bond shift; the two alternatives can be readily distinguished.) By this methodology the course of the cyclization was shown to be inversion [167, 168].

The stereochemistry of the reverse reaction, opening of cAMP and AMP by beef heart cyclic CMP phosphodiesterase was found by Gerlt and coworkers to involve inversion in the deoxyadenosine series [169], analogous to the corresponding opening of the cyclic phosphothioate cAMPS (Fig. 66). The same point was demonstrated in the adenosine series [168, 170] by opening the stereospecifically labeled cAMP, formed as shown in Fig. 75, with $H_2{}^{17}O$ to form stereospecifically labeled AMP (*122*, Fig. 75). That this material had indeed the same stereochemistry as *122* was established by recyclizing it as shown in Fig. 75. Once again, compounds *123* and *124* were obtained predominantly (a considerable amount of isotope dilution occurs in the experiment) and it is therefore clear that the sequence of ring opening and ring closure (cAMP→AMP→cAMP) goes with overall retention of configuration. Thus, since the second step involves inversion, the first step must do so also.[47]

Inversion of configuration at phosphorus occurs in all the kinase mediated reactions studied so far [165, 173]. In contrast, J. P. Knowles' group has shown that alkaline phosphatase [174] and phosphoglycerate mutases [175] induce retention of configuration, presumably as a result of a two-step process, each step involving inversion, although an "adjacent mechanism" with pseudorotation at phosphorus is an alternative possibility. The method of isotopic stereochemical analysis used in the work of Knowles' group [164, 166, 173–177] was mass spectrometry involving metastable ions [163, 176], though in later work [174] they have also described an NMR method of stereochemical analysis.

6 Prochirality and Two-dimensional Chirality

We conclude with an alternative view of prochirality [11] which is based on chirality in two dimensions. A scalene triangle, or any triangle with three differently labeled vertices A, B, C exists in two-dimensional mirror images, as shown in Fig. 76.

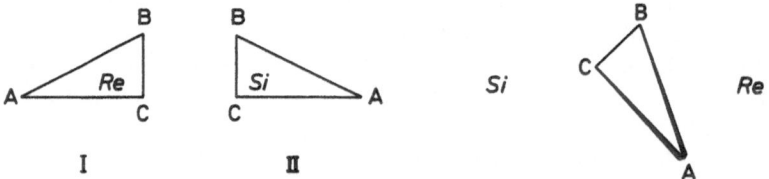

Fig. 76. Two-dimensional chirality

---

[47] Because of the earlier-mentioned error in configurational assignment of *120* (Fig. 74), the initial conclusion was that both steps involve retention and that the phosphate and phosphothioate reactions differ in stereochemical course [170]. This conclusion was subsequently revised [168] as explained above (see also footnote 25 in Ref. 169).

These mirror image representations cannot be made to coincide as long as they are maintained in the plane of the paper. As we have already seen (Sect. 3), such two-dimensionally chiral representations may be taken to depict heterotopic faces. If the sequence is A > B > C, triangle I represents the *Re* face and triangle II the *Si* face.

But if one now proceeds into three-dimensional space, one may say that the plane of triangle I, say, divides all space into two halves, one in front of triangle I (*Re* space) the other behind (*Si* space). On the right of Fig. 76 a sideways view of the triangle is given with the *Re* space to the right and the *Si* space to the left. Three-dimensional prochirality may now be considered in the following terms: A ligand identical with one already present (say A) is made to form a bond from either the *Re* or the *Si* side of the triangle ABC. This ligand A' thus sees mirror-image representations of ABC, depending on the side where it is placed. The ligand may now be labeled A_{Re} or A_{Si}, depending on the halfspace in which it finds itself, as shown in Fig. 77. (A, B, C and A' may be considered to be attached to a — three-dimensional — prochiral center.)

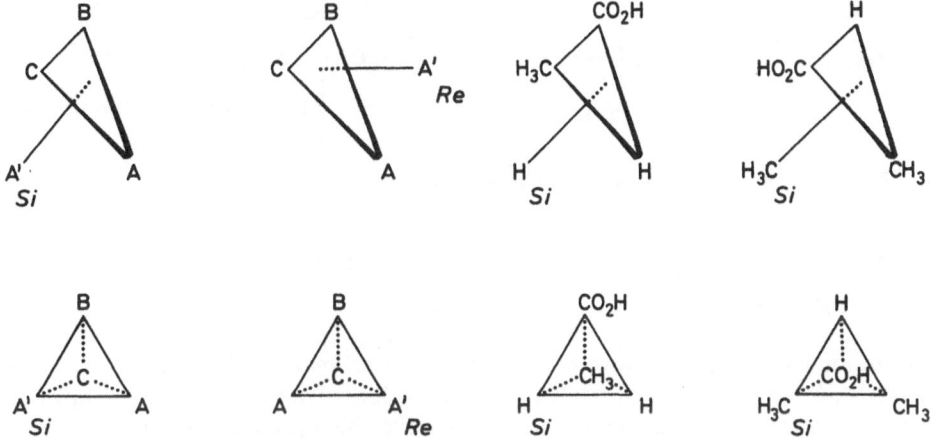

Fig. 77. Ligand attached to prochiral center

The representation in Fig. 77 has several advantages. It closely correlates the prochirality of the plane with the prochirality of the chiral center formed by addition to that plane. Even the symbols correspond. And the symbolism is equally applicable to addition to various types of planes, such as ABC, ABF, AFH, etc.

There is, however, one serious disadvantage: the symbolism implied in Fig. 77 leads to a loss of the relationship of symbols as between prochirality and chirality. We had pointed out, earlier, that since all stable isotopes in practical use are heavier than the ordinary nuclides, replacing a *pro-R* ligand by an isotopically labeled one will give a molecule of *R*-chirality, and similar for *pro-S*; this relationship was extensively used in Sec. 5 (e.g. Fig. 55). Unfortunately such a relationship no longer necessarily holds in the alternative nomenclature, as shown in Fig. 78.

Ernest L. Eliel

If (Fig. 77) the homomorphic ligands are A or C in the sequence (as in $C_6H_5\overset{O}{\underset{O}{\overset{\|}{\underset{\|}{S}}}}CH_3$

or $CH_3CH_2CO_2H$ — Fig. 78) A_{Re} and C_{Re} correspond to A_R and C_R (and similarly for A_{Si}, C_{Si}, A_S, C_S) but for the intermediate ligand B the situation is otherwise; B_{Re} corresponds to B_S and B_{Si} to B_R: cf. the case of isobutyric acid in Fig. 78. For this reason among others the alternative view has not been widely espoused, although it has been used in some recent publications [138,139].

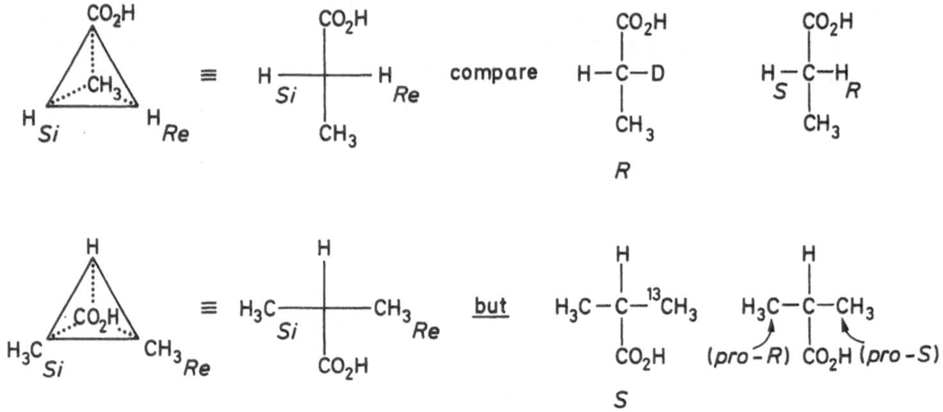

Fig. 78. Alternative nomenclature for prochiral ligands

7 Acknowledgements

The author is grateful to Professors H. Floss, H. Hirschmann, J. Knowles, D. Arigoni and K. Mislow and Dr. K. Hanson for a careful reading of this chapter and a number of helpful comments and suggestions. I wish to take this opportunity to express may special appreciation to Professor Mislow for having introduced me to the topic of this article in the mid-1960's (refs. 5 and 25) and for numerous stimulating and incisive exchanges of ideas regarding concepts and terminology extending over many years. I am indebted to the John Simon Guggenheim Foundation for a fellowship during the tenure of which the first draft of this chapter was written.

8 References

1 a. Cahn, R. S., Ingold, C. K., Prelog, V.: Angew. Chem. Int. Ed. (Engl.) 5, 385 (1966); b. Prelog, V., Helmchen, G.: *ibid*, in press
2. see, however, Hirschmann, H., Hanson, K.: Topics in Stereochemistry, Vol. 14, in press for a discussion of the adequacy of these terms
3. Hirschmann, H., Hanson, K. R.: Tetrahedron 30, 3649 (1974)
4. Hanson, K. R.: J. Am. Chem. Soc. 88, 2731 (1966)
5. Mislow, K., Raban, M.: Topics in Stereochemistry 1, 1 (1967)

6. Hirschmann, H., Hanson, K. R.: Eur. J. Biochem. *22*, 301 (1971)
7. See also: Eliel, E. L.: J. Chem. Educ. *57*, 52 (1980)
8. McCasland, G. E.: A New General System for the Naming of Stereoisomers: Chemical Abstracts, Columbus, OH, 1950
9. Hirschmann, H., Hanson, K. R.: J. Org. Chem. *36*, 3293 (1971)
10. cf. Eliel, E. L.: Stereochemistry of Carbon Compounds, New York: McGraw-Hill 1962. Regarding some question as to the definition of this term, cf. refs. 3 11
11. Prelog, V., Helmchen, G.: Helv. Chim. Acta *55*, 2581 (1972)
12. Cram, D. J. et al.: Tetrahedron *30*, 1757 (1974)
13. Marckwald, W.: Ber. *37*, 349 (1904)
14. Evans, E. A., Slotin, L.: J. Biol. Chem. *141*, 439 (1941)
    Wood, H. G. et al.: J. Biol. Chem. *142*, 31 (1942)
15. Bublitz, C., Kennedy, E. P.: J. Biol. Chem. *211*, 951 (1954)
16a. Ogston, A. G.: Nature *162*, 963 (1948). For a historical review, see: Bentley, R.: Nature 276, 673 (1978). b. Easson, L. H., Stedman, E.: Biochem. J. *27*, 1257 (1933). c. Bergmann, M.: Science, *79*, 439 (1934)
17. Schwartz, P., Carter, H. E.: Proc. Natl. Acad. Sci. U.S. *40*, 499 (1954)
18. Hirschmann, H.: Newer Aspects of Enzymatic Stereochemistry, in: Comprehensive Biochemistry, Vol. 12, pp. 236ff. (Florkin, M., Stotz, G. H. eds). New York: American Elsevier Publishing Co. 1964
19. Jones, J. B.: Stereochemical Considerations and Terminologies of Biochemical Importance, in: Applications of Biochemical Systems in Organic Chemistry, Part 1, p. 479. (Jones, J. B., Sih, C. J., Perlman, D. eds.). New York: Wiley 1976
20. Drysdale, J. J., Phillips, W. D.: J. Am. Chem. Soc. *79*, 319 (1957)
21. Nair, P. M., Roberts, J. D.: J. Am. Chem. Soc. *79*, 4565 (1957)
22. Waugh, J. S., Cotton, F. A.: J. Phys. Chem. *65*, 562 (1961)
23. Gutowsky, H. S.: J. Chem. Phys. *37*, 2196 (1962)
24. Kaloustian, S. A., Kaloustian, M. K.: J. Chem. Educ. *52*, 56 (1975)
25. Mislow, K.: Introduction to Stereochemistry, New York: W. A. Benjamin, Inc., 1965, p. 73
26a. Pirkle, W. H., Beare, S. D., Muntz, R. L.: J. Am. Chem. Soc. *91*, 4575 (1969)
   b. Pirkle, W. H., Muntz, R. L., Paul, I. C.: J. Am. Chem. Soc. *93*, 2817 (1971)
   c. Kainosho, M. et al.: J. Am. Chem. Soc. *94*, 5924 (1972)
27. For a review see: Pirkle, W. H., Hoover, D. J.: Topics in Stereochemistry, *13*, 263 (1982)
28. Morrison, J. D., Mosher, H. S.: Asymmetric Organic Reactions, Englewood Cliffs, N. J.: Prentice-Hall 1971
29a. Kagan, H. B., Fiaud, J. C.: Topics in Stereochemistry *10*, 175 (1978)
   b. Valentine, Jr., D., Scott, J. W.: Synthesis, 329 (1978)
   c. ApSimon, J. W., Seguin, R. P.: Tetrahedron *35*, 2797 (1979)
30. Reisse, J. et al.: J. Am. Chem. Soc. *100*, 911 (1978)
31. Eliel, E. L.: J. Chem. Educ. *48*, 163 (1971)
32. Hanson, K. R.: Ann. Rev. Biochem. *45*, 307 (1976)
33. Martin, M. L., Mantione, R., Martin, G. J.: Tetrahedron Lett. 3185 (1965)
34. Martin, M. L., Martin, G. J., Couffignal, R.: J. Chem. Soc. (B) 1282 (1971)
35. Beaulieu, P. L., Morisset, V. M., Garratt, D. G.: Can. J. Chem. *58*, 928 (1980)
36. Slocum, D. W., Stonemark, F.: Tetrahedron Lett. 3291 (1971)
37. Fieser, L. F., Fieser, M.: Steroids, New York: Reinhold 1959
38. Rose, I. A. et al.: Proc. Natl. Acad. Sci. USA 77, 2439 (1980)
39. Perutz, M. F. et al.: Nature (London) *185*, 416 (1960); Kendrew, J. C. et al.: ibid. *185*, 422 (1960)
40. Jennings, W. B.: Chem. Rev. *75*, 307 (1975)
41. Siddall, T. H., Stewart, W. E.: Progr. Nucl. Mag. Resonance Spectros. *5*, 33 (1969)
42. Sokolov, V. I., Petrovskii, P. V., Reutov, O. A.: J. Organometal. Chem. *59*, C27 (1973)
43. Whitesides, G. M., Holtz, D., Roberts, J. D.: J. Am. Chem. Soc. *86*, 2628 (1964)
44. Schiemenz, G. P., Rast, H.: Tetrahedron Lett. 4685 (1971)
45. Martin, M. L., Martin, G. J.: Bull. Soc. Chim. Fr. 2117 (1966)
46. Devriese, G. et al.: Bull. Soc. Chim. Belg. *85*, 167 (1976)
47. Boucher, H., Bosnich, B.: J. Am. Chem. Soc. *99*, 6253 (1977)

48. Schurig, V.: Tetrahedron Lett. 3977 (1977)
49a. Martin, M. L., Mantione, R., Martin, G. J.: Tetrahedron Lett. 3873 (1966); *id. ibid*: 4809 (1967); b. Musierowicz, S., Wroblewski, A. E.: Tetrahedron *36*, 1375 (1980)
50. cf. Wilson, N. K., Stothers, J. B.: Topics in Stereochemistry *8*, 1 (1974), expecially p. 17
51. Goering, H. L. et al.: J. Am. Chem. Soc. *96*, 1493 (1974)
52. Fraser, R. R., Petit, M. A., Miskow, M.: J. Am. Chem. Soc. *94*, 3253 (1972)
53a. Whitesides, G. M., Lewis, D. W.: J. Am. Chem. Soc. *92*, 6979 (1970)
   b. *Id., ibid., 93*, 5914 (1971)
   c. Goering, H. L., Eikenberry, J. N., Koermer, G. S.: *ibid, 93*, 5913 (1971)
   d. Fraser, R. R., Petit, M. A., Saunders, J. K.: J. Chem. Soc. Chem. Comm. 1450 (1971)
   e. Sullivan, G.: Topics in Stereochemistry *10*, 287 (1978)
54. cf. Binsch, G.: Topics in Stereochemistry *3*, 97 (1968), p. 158
55. Fraser, R. R., Schuber, F. J.: Can. J. Chem. *48*, 633 (1970); Fraser, R. R., Schuber, F. J., Wigfield, Y. Y.: J. Am. Chem. Soc. *94*, 8795 (1972)
56. Brink, M., Schjånberg, E.: J. Prak. Chem. *322*, 685 (1980)
57. Biernacki, W., Dabrowski, J., Ejchart, A.: Org. Magn. Reson. *4*, 443 (1972)
58. cf. Gaudemer, A.: in: Stereochemistry, Fundamentals and Methods, Vol. 1, p. 73. (Kagan, H. B. ed.). Stuttgart: Georg Thieme 1977
59. Giardina', D. et al.: Tetrahedron *35*, 249 (1979)
60. Giardina', D. et al.: Tetrahedron *36*, 3565 (1980)
61. Hill, R. K., Chan, T.-H.: Tetrahedron *21*, 2015 (1965)
62a. Kost, D., Raban, M.: J. Am. Chem. Soc. *94*, 2533 (1972)
   b. Raban, M., Lauderback, S. K., Kost, D.: *ibid., 97*, 5178 (1975)
63a. Raban, M., Mislow, K.: Tetrahedron Lett. 3961 (1966). b) Mislow, K., O'Brien, R. E., Schaefer, H.: J. Am. Chem. Soc. *82*, 5512 (1960)
64. Reuben, J.: J. Am. Chem. Soc. *102*, 2232 (1980)
65a. Hill, R. K., Yan, S., Arfin, S. M.: J. Am. Chem. Soc. *95*, 7857 (1973)
   b. Aberhart, D. J., Lin, L. J.: J. Am. Chem. Soc. *95*, 7859 (1973); *id*., J. Chem. Soc. Perkin I, 2320 (1974)
   c. Baldwin, J. E. et al.: J. Am. Chem. Soc. *95*, 3796 (1973)
   d. Kluender, H. et al.: J. Am. Chem. Soc. *95*, 6149 (1973)
   e. See also Townsend, C. A., Neese, A. S., Theis, H. B.: J. Chem. Soc. Chem. Comm. 116 (1982)
66a. Gerlach, H., Zagalak, B.: J. Chem. Soc. Chem. Comm., 274 (1973)
66a. Battersby, A. R., Staunton, J.: Tetrahedron *30*, 1707 (1974)
67. Kobayashi, K., Sugawara, T., Iwamura, H.: J. Chem. Soc. Chem. Comm. 479 (1981)
68. Tabacik, V.: Tetrahedron Lett. 555, 561 (1968)
69. Raban, M.: Tetrahedron Lett. 3105 (1966)
70. Newmark, R. A., Sederholm, C. H.: J. Chem. Phys. *43*, 602 (1965). See also: Binsch, G.: J. Am. Chem. Soc. *95*, 190 (1973); Norris, R. D., Binsch, G.: *ibid., 95*, 182 (1973)
71a. Binsch, G., Franzen, G. R.: J. Am. Chem. Soc. *91*, 3999 (1969)
   b. *Id., ibid., 95*, 175 (1973)
72. McKenna, J., McKenna, J. M., Wesby, B. A.: Chem. Commun. *867* (1970)
73. Gielen, M., Close, V., de Poorter, B.: Bull. Soc. Chim. Belges *83*, 339 (1974)
74. Morris, D. G. et al.: Tetrahedron Lett. 3179 (1973)
75. Lambert, J. B.: Topics in Stereochemistry *6*, 19 (1971)
76. Lehn, J.-M.: Topics in Current Chem. *15*, 311 (1970)
77. Kessler, H.: Angew. Chem. Int. Ed. Engl. *9*, 219 (1970)
78. Internal Rotation in Molecules, (Orville-Thomas, W. J. ed.). New York: John Wiley & Sons 1974
79. Dynamic Nuclear Magnetic Resonance Spectroscopy, (Jackman, L. M., Cotton, F. A. eds.). New York: Academic Press 1975
80. Steigel, A. in: Dynamic NMR Spectroscopy, NMR 15, (Diehl, P., Fluck, E., Kosfeld, R. eds.). Heidelberg: Springer-Verlag 1978
81. Roberts, J. D.: Pure Appl. Chem. *51*, 1037 (1979)
82. Binsch, G., Kessler, H.: Angew. Chem. Int. Ed. Engl. *19*, 411 (1980)
83. Leonard, J. E., Hammond, G. S., Simmons, H. E.: J. Am. Chem. Soc. *97*, 5052 (1975)
84. Spassov, S. L. et al.: J. Am. Chem. Soc. *89*, 88 (1967)

85. Roberts, J. D.: Angew. Chem. Int. Ed. Engl. *2*, 53 (1963)
86. Binsch, G.: Band-Shape Analysis, in: ref. 79
87. Pople, J. A., Schneider, W. G., Bernstein, H. J.: High-Resolution Nuclear Magnetic Resonance. New York: McGraw-Hill 1959
88. Kurland, R. J., Rubin, M. B., Wise, W. B.: J. Chem. Phys. *40*, 2426 (1964)
89. Kost, D., Carlson, E. H., Raban, M.: Chem. Commun. *656* (1971)
90. Shanan-Atidi, H., Bar-Eli, K. H.: J. Phys. Chem. *74*, 961 (1970)
91. Dahlqvist, K.-I., Forsén, S.: J. Phys. Chem. *69*, 4062 (1965)
92. Binsch, G., Eliel, E. L., Kessler, H.: Angew. Chem. Int. Ed. Engl. *10*, 570 (1971)
93. Balaban, A. T., Farcasiu, D.: J. Am. Chem. Soc. *89*, 1958 (1967)
94. Sternhell, S.: Rotation about Single Bonds in Organic Molecules, in: ref. 79
95. Jackman, L. M.: Rotation about Partial Double Bonds in Organic Molecules, in: ref. 79
96. Kalinowski, H.-O., Kessler, H.: Topics Stereochem. *7*, 295 (1973)
97. Weigert, F. J., Mahler, W.: J. Am. Chem. Soc. *94*, 5314 (1972)
98. Bushweller, C. H. et al.: J. Am. Chem. Soc. *94*, 4743 (1972)
99. Jennings, W. B. et al.: Org. Magn. Resonance *9*, 151 (1977)
100. Bushweller, C. H. et al.: J. Am. Chem. Soc. *97*, 4338 (1975)
101. Saunders, M., Yamada, F.: J. Am. Chem. Soc. *85*, 1882 (1963)
102. Laszlo, P.: cited in: ref. 92
103. Rattet, L. S., Mandell, L., Goldstein, J. H.: J. Am. Chem. Soc. *89*, 2253 (1967)
104. McFarlane, W.: Chem. Commun. 229 (1968)
105. For an alternative notation, see: Haigh, C. W.: J. Chem. Soc. (A) 1682 (1970)
106. Mislow, K.: personal communication
107. Izumi, Y.: Angew. Chem. Int. Ed. Engl. *10*, 871 (1971)
108. Izumi, Y., Tai, A.: Stereodifferentiating Reactions, New York: Academic Press 1977
109. Eliel, E. L., Hartmann, A. A., Abatjoglou, A. G.: J. Am. Chem. Soc. *96*, 1807 (1974)
110. Sicher, J., Zavada, J., Pankova, M.: Coll. Czech. Chem. Comm. *36*, 3140 (1971) and earlier papers
111. Arigoni, D., Eliel, E. L.: Topics Stereochem. *4*, 127 (1969)
112. Verbit, L.: Progr. Phys. Org. Chem. *7*, 51 (1970)
113. Hanson, K. R.: Ann. Rev. Plant Physiol. *23*, 335 (1972)
114. Bentley, R.: The Use of Biochemical Methods for Determination of Configuration in: Applications of Biochemical Systems in Organic Chemistry, Part 1, p. 403. (Jones, J. B., Sih, C. J., Perlman, D. eds.) New York, Wiley 1976
115. Hill, R. K.: Enzymatic Stereospecificity at Prochiral Centers of Amino Acids, in: Bio-organic chemistry, Vol. 2, p. 111. (van Tamelen, E. E. ed.). New York: Academic Press 1978
116. Simon, H., Kraus, A.: Hydrogen Isotope Transfer in Biological Processes, in: Isotopes in Organic Chemistry, Vol. 2, p. 153. (Buncel, E., Lee, C. C. eds.). Amsterdam: Elsevier 1976
117. Young, D. W.: Stereospecific Synthesis of Tritium Labelled Organic Compounds using Chemical and Biological Methods, Vol. 4, p. 177, *ibid.* 1978
118. Cane, D. E.: Tetrahedron *36*, 1109 (1980)
119. Bentley, R.: Molecular Asymmetry in Biology, Vol. 1. New York: Academic Press 1969; Vol. 2, 1970
120. Alworth, W. L.: Stereochemistry and its Applications in Biochemistry, New York: John Wiley & Sons 1972; 2nd ed. in press 1982
121. Loewus, F. A., Westheimer, F. H., Vennesland, B.: J. Am. Chem. Soc. *75*, 5018 (1953)
122. Levy, H. R., Loewus, F. A., Vennesland, B.: J. Am. Chem. Soc. *79*, 2949 (1957)
123. Hanson, K. R., Rose, I. A.: Acc'ts Chem. Res. *8*, 1 (1975)
124. Vennesland, B.: Stereospecificity in Biology, in: Topics in Current Chemistry *48*, 39 (1974); Adams, M. J. et al.: Proc. Nat. Acad. Sci. USA *70*, 1968 (1973)
125. Cornforth, J. W. et al. Biochem. Biophys. Res. Commun. *9*, 371 (1962)
126. cf. You, K.-S. et al.: Trends in Biochemical Sciences *3*, 265 (1978)
127. Englard, S., Colowick, S. P.: J. Biol. Chem. *221*, 1019 (1956)
128. Krasna, A. I.: J. Biol. Chem. *233*, 1010 (1958)
129. Chibata, S.: Applications of Immobilized Enzymes for Asymmetric Reactions, Am. Chem. Soc. Symp. Series 185, Washington, DC, 1982

130. Gawron, O., Fondy, T. P.: J. Am. Chem. Soc. *81*, 6333 (1959); Gawron, O., Glaid, A. J., Fondy, T. P.; *ibid., 83*, 3634 (1961)
131. Anet, F. A. L.: J. Am. Chem. Soc. *82*, 994 (1960)
132. England, S., Colowick, S. P.: J. Biol. Chem. *226*, 1047 (1957)
133. Hanson, K. R., Rose, I. A.: Proc. Natl. Acadm. Sci. US *50*, 981 (1963)
134. cf. Ref. 111, p. 193
135. England, S.: J. Biol. Chem. *235*, 1510 (1960)
136. Gottschalk, G., Barker, H. A.: Biochemistry *5*, 1125 (1966), *6*, 1027 (1967)
137. Cornforth, J. W.: J. Mol. Catalysis *1*, 145 (1976); ib., Quart. Rev. *23*, 125 (1969); id., Chem. Soc. Rev. *2*, 1 (1973)
138. Battersby, A. R., Chrystal, E. J. T., Staunton, J.: J. Chem. Soc. Perkin I, 31 (1980)
139. Battersby, A. R. et al.: J. Chem. Soc. Perkin I, 43 (1980)
140. Chang, M. N. T., Walsh, C.: J. Am. Chem. Soc. *102*, 7368 (1980)
141. Neuss, N. et al.: J. Am. Chem. Soc. *95*, 3797 (1973); see also ref. 65c
142. Archer, R. A. et al.: Chem. Comm. 1291 (1970)
143. Burgers, P. M. J. et al.: J. Biol. Chem. *254*, 9959 (1979)
144. Baraniak, J. et al.: J.C.S. Chem. Comm. 940 (1979)
145a. Burgers, P. M. J., Eckstein, F.: Proc. Natl. Acad. Sci. USA *75*, 4798 (1978); b. Jarvest, R. L., Lowe, G.: J.C.S. Chem. Comm. 364 (1979)
146. Cohn, M., Hu, A.: Proc. Natl. Acad. Sci. USA *75*, 200 (1978). See also Lowe, G., Sproat, B. S.: J.C.S. Chem. Comm. 565 (1978)
147. Gerlt, J. A., Coderre, J. A., Wolin, M. S.: J. Biol. Chem. *255*, 331 (1980)
148. cf. Ref. 120, p. 108
149. For a review: See Floss, H. G., Tsai, M.-D.: Adv. Enzymology *50*, 243 (1979)
150a. Cornforth, J. W. et al.: Nature *221*, 1212 (1969); *ibid.*, Eur. J. Biochem. *14*, 1 (1970); b. Lüthy, J., Rétey, J., Arigoni, D.: Nature 221, 1213 (1969)
151. Lenz, H. et al.: Eur. J. Biochem. *24*, 207 (1971)
152. Townsend, C. A., Scholl, T., Arigoni, D.: J.C.S. Chem. Comm. 921 (1975)
153. Woodard, R. W. et al.: J. Am. Chem. Soc. *102*, 6314 (1980)
154. Altman, L. J. et al.: J. Am. Chem. Soc. *100*, 3235 (1978)
155a. Lenz, H. et al.: Z. Physiol. Chem. *352*, 517 (1971)
   b. Lenz, H., Eggerer, H.: Eur. J. Biochem. *65*, 237 (1976)
156. Eggerer, H. et al.: Nature *226*, 517 (1970)
157. Rétey, J., Lüthy, J., Arigoni, D.: Nature *226*, 519 (1970)
158. Sedgwick, B., Cornforth, J. W.: Eur. J. Biochem. *75*, 465 (1977)
159. Blättler, W. A.: Ph. D. Dissertation No. 6127, ETH, Zürich 1978
160. Caspi, E., Piper, J., Shapiro, S.: J. Chem. Soc. Chem. Comm. 76 (1981)
161. Woodward, R. W. et al.: J. Biol. Chem. *255*, 9124 (1980)
162. Woodward, R. W., Weaver, J., Floss, H. G.: Arch. Biochem. Biophys. *207*, 51 (1981)
163. Abbott, S. J. et al.: J. Am. Chem. Soc. *100*, 2558 (1978)
164. Cullis, P. M., Lowe, G.: J. Chem. Soc. Chem. Comm. 512 (1978)
165. Lowe, G. et al.: Phil. Trans. R. Soy Lond. B. *293*, 75 (1981)
166. See also: Knowles, J. R.: Ann. Rev. Biochem. *49*, 877 (1980)
167. Jarvest, R. L., Lowe, G., Potter, B. V. L.: J. Chem. Soc. Chem. Comm. 1142 (1980)
168. Cullis, P. M. et al.: J. Chem. Soc. Chem. Comm. 245 (1981)
169. Coderre, J. A., Mehdi, S., Gerlt, J. A.: J. Am. Chem. Soc. *103*, 1872 (1981)
170. Jarvest, R. L., Lowe, G.: J. Chem. Soc. Chem. Comm. 1145 (1980)
171a. Tsai, M. D.: Biochem. *18*, 1468 (1979)
   b. See also: Lowe, G. et al.: J. Chem. Soc. Chem. Comm. 733 (1979)
172. Gerlt, J. A., Coderre, J. A.: J. Am. Chem. Soc. *102*, 4531 (1980)
173. Blättler, W. A., Knowles, J. R.: J. Am. Chem. Soc. *101*, 510 (1979)
174. Jones, S. R., Kindman, L. A., Knowles, J. R.: Nature *275*, 564 (1978)
175. Blättler, W. A., Knowles, J. R.: Biochemistry *19*, 738 (1980)
176. Abbott, S. J. et al.: J. Am. Chem. Soc. *101*, 4323 (1979)
177. Buchwald, S. L., Knowles, J. R.: J. Am. Chem. Soc. *102*, 6601 (1980)
178. Prelog, V.: personal communication.

Asymmetric Hydroformylation

Giambattista Consiglio and Piero Pino

Eidgenössische Technische Hochschule, Technisch-Chemisches Laboratorium,
ETH-Zentrum, CH-8092 Zürich, Switzerland

Table of Contents

Abbreviations

 1) R*-SAL   = N-(1-phenylethyl)salicylaldimine
 2) DIOP   = 2,2-dimethyl-4,5-bis(diphenylphosphinomethyl)-1,3-dioxolane
 3) DIOP-DBP   = 2,2-dimethyl-4,5-bis(5H-dibenzophosphol-5-ylmethyl)-
                  -1,3-dioxolane
 4) CHIRAPHOS   = 2,3-bis(diphenylphosphino)butane
 5) CBDPP   = 1,2-bis(diphenylphosphinomethyl)cyclobutane
 6) CBDBP   1,2-bis(5 H-dibenzophosphol-5-ylmethyl)cyclobutane
 7) CHDPP   = 1,2-bis(diphenylphosphinomethyl)cyclohexane
 8) CHDBP   = 1,2-bis(5H-dibenzophosphol-5-ylmethyl)cyclohexane
 9) CHDPPO   = 1,2-bis(diphenylphosphinoxy)cyclohexane
10) BzMePhP*   = benzyl-methyl-phenylphosphine
11) CAMP   = cyclohexyl-o-anisyl-methylphosphine
12) M(2MB)PP   = methyl(2-methylbutyl)phenylphosphine
13) MePhPr^{n}P* = methyl-phenyl-n-propylphosphine
14) NMDPP   = neomenthyldiphenylphosphine
15) PAMP   = phenyl-o-anisyl-methylphosphine

1 Introduction

Hydroformylation of olefins [1,2] was discovered by Roelen in 1938 and was called "oxo" reaction. It is of paramount importance to the chemical industry and its mechanism is still the subject of numerous investigations [3]. Asymmetric hydroformylation is in principle a convenient synthetic method for obtaining in one step optically active aldehydes from olefinic substrates.

The first positive results [4] were obtained at the end of 1971 by using the rhodium complexes in the presence of optically active phosphines [5] or dicobaltoctacarbonyl as the catalyst precursor in the presence of optically active N-alkylsalicylaldimines [6].

After the first successful asymmetric hydroformylation, although in low optical yields, the reaction was further investigated by different groups all over the world. The results have been rather disappointing from a synthetic point of view, as in a few cases only, optical yields as high as 30 to 50% have been achieved. However, some interesting information has been obtained, both on the mechanism of hydroformylation and on the basic aspects of homogeneous asymmetric catalysis by transition metal complexes.

In this review the synthetic aspects of asymmetric hydroformylation will be discussed first; the experimental data relevant to attempt a rationalization of the results will then be considered. The closely related synthesis of optically active aldehydes by hydroformylation of optically active olefinic substrates in the presence of achiral catalysts [7,8] and the different asymmetric hydrocarbonylation reactions, such as the synthesis of esters from olefins, carbon monoxide and alcohols in the presence of optically active catalysts [9], are beyond the scope of this review and will not be discussed here.

2 Formation of Optically Active Aldehydes in Olefin Hydroformylation

Optically active aldehydes can be obtained by asymmetric hydroformylation of olefinic substrates when at least one asymmetric carbon atom is formed either by addition of a formyl group or of a hydrogen atom to an unsaturated carbon atom (Scheme 1, reactions (1) and (2)). In the case of trisubstituted olefins, two new asymmetric carbon atoms can form; due to the *cis* stereochemistry of the reaction [10], in the absence of isomerization, the formation of only one epimer is expected.

Reactions (1) and (2) in Scheme 1 are enantioface-discriminating[1] reactions, according to the nomenclature proposed by Izumi and Tai [11].

The hydroformylation of racemic olefins presents a further possibility of obtaining optically active aldehydes; in the presence of a chiral non-racemic[2] catalyst, a racemate resolution [13] takes place and both an optically active aldehyde and an optically active non-reacted substrate are generally produced if conversion is kept below 100%. This type of asymmetric hydroformylation can be classified as an

---

1 In the cases in which two isomeric products are formed, enantioface differentiation is not necessary to achieve asymmetric hydroformylation [12] (see Sect. 2.1.5.).

2 In this review we shall use for simplicity the expression "chiral" to mean chiral non-racemic

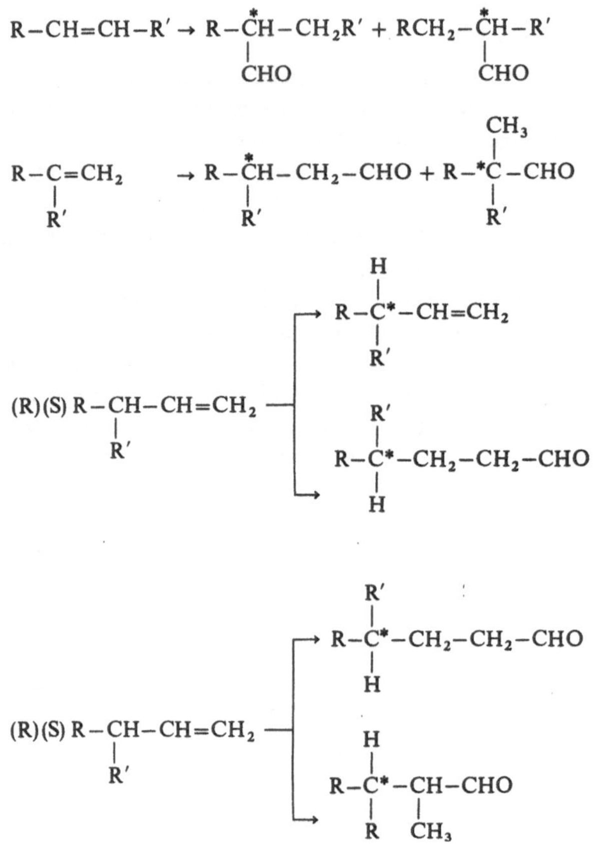

$$R-CH=CH-R' \rightarrow R-\overset{*}{C}H-CH_2R' + RCH_2-\overset{*}{C}H-R' \qquad (1)$$
$$\qquad\qquad\qquad\quad | \qquad\qquad\qquad\qquad |$$
$$\qquad\qquad\qquad\quad CHO \qquad\qquad\qquad\quad CHO$$

$$R-C=CH_2 \quad \rightarrow R-\overset{*}{C}H-CH_2-CHO + R-\overset{*}{C}-CHO \qquad (2)$$

where the right product bears a CH_3 group.

$$(R)(S)\ R-CH-CH=CH_2 \longrightarrow \qquad (3)$$

with products:

$$R-\overset{H}{\underset{R'}{C^*}}-CH=CH_2$$

$$R-\overset{R'}{\underset{H}{C^*}}-CH_2-CH_2-CHO$$

$$(R)(S)\ R-CH-CH=CH_2 \longrightarrow \qquad (4)$$

with products:

$$R-\overset{R'}{\underset{H}{C^*}}-CH_2-CH_2-CHO$$

$$R-\overset{H}{\underset{R}{C^*}}-CH-CHO$$
with a CH_3 substituent.

Scheme 1

enantiomer-differentiating synthesis [11] (Scheme 1, reaction 3) accompanied by a kinetic resolution of the substrate.

If during the hydroformylation of a racemic olefin more than one aldehyde is formed, an enantiomer-differentiating synthesis with formation of two (or more) optically active aldehydes can be obtained [14], even if the conversion is complete (Scheme 1, reaction 4, and Scheme 2).

Asymmetric hydroformylations of all the above types have been achieved with rhodium catalysts; enantioface- and enantiomer-discriminating hydroformylations also occur with cobalt and platinum catalysts whereas with ruthenium or iridium complexes only enantioface-discriminating synthesis has been reported up to now (see Sect. 2.1.4.).

Complexes containing chiral ligands have been used as precursors of the asymmetric hydroformylation catalysts. However, in most cases, the complexes used contain achiral ligands and the chiral ligand is added to the reaction medium either before or during the reaction. It is, in general, assumed that the chiral catalyst forms "in situ" by exchange between the non-chiral and the chiral ligand.

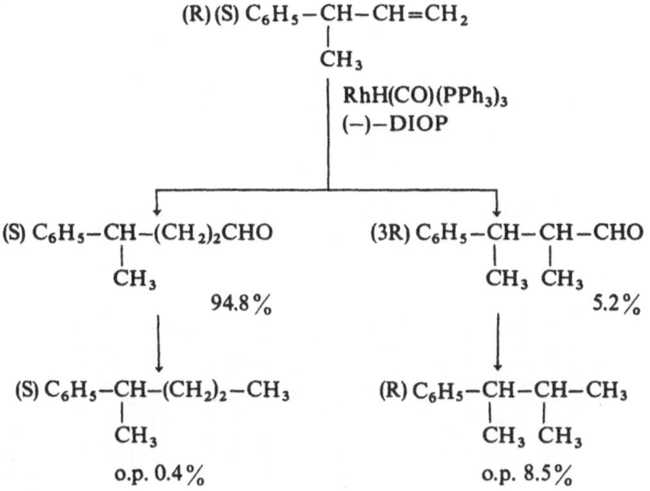

(R)(S) $C_6H_5-CH-CH=CH_2$
|
CH_3

$RhH(CO)(PPh_3)_3$
$(-)-DIOP$

(S) $C_6H_5-CH-(CH_2)_2CHO$
|
CH_3
94.8%

(3R) $C_6H_5-CH-CH-CHO$
| |
CH_3 CH_3
5.2%

(S) $C_6H_5-CH-(CH_2)_2-CH_3$
|
CH_3
o.p. 0.4%

(R) $C_6H_5-CH-CH-CH_3$
| |
CH_3 CH_3
o.p. 8.5%

Scheme 2

However, this is not the only possibility of obtaining a chiral catalyst in the presence of a chiral ligand [55].

2.1 Enantioface-Discriminating Hydroformylation

The experimental data given in the following sections are ordered according to the metal present in the catalytic complexes. The optical purity of the ligands used is reported in the appendix.

2.1.1 Cobalt-Catalyzed Hydroformylation

No substantial progress has been made in the field of cobalt-catalyzed asymmetric hydroformylation since our last review on this subject [15]. Besides (+)-N-(1-phenyl-ethyl)salicylaldimine, which was originally used as asymmetric ligand [6], a chiral catalyst formed "in situ" from $HCo(CO)_4$ and $(-)$-DIOP has been employed [16] (Table 1). With the latter catalytic system, optical yields of 2.7% and 1.2% have been obtained in the case of (Z)-2-butene and of bicyclo[2,2,2]oct-2-ene, respectively.

(+) (S)-3-sec-Butylpyridine, which interacts with the cobalt catalytic system (as shown by the increase of the reaction rate), does not cause asymmetric induction when styrene is used as the substrate [17]. Alkoxyalkylidenetricobaltmonocarbonyl cluster complexes bearing chiral alkoxy groups used as catalyst precursors do not give optically active aldehydes either [18].

The only substrate which has been hydroformylated using chiral cobalt catalysts with an optical yield comparable to those obtained with other metals is styrene (Table 1); in fact, in this case, optical yields up to 15% were obtained working in the presence of ethyl orthoformate [15] to avoid racemization of 2-phenylpropanal. The changes in the prevailing absolute configuration of the synthesized aldehyde observed both in the styrene and 2-phenyl-1-propene hydroformylation upon

Table 1. Cobalt-catalyzed asymmetric hydroformylation[a]

Substrate	Chiral ligand	Metal-to-ligand molar ratio	Reaction conditions: t °C	Reaction conditions: P_{H_2} (atm)	Reaction conditions: P_{CO} (atm)	Reaction conditions: time (hrs)	Conversion[b]	Yield[c]	Hydroformylation product and isomeric composition[d]		Chiral reaction product: o.p. %	Chiral reaction product: abs. conf.	Ref.
styrene	(+)-R*-Sal	1/8	120	40	40	2.5	98	46	3-phenylpropanal	41			
									2-phenylpropanal	59	1.9	(S)	6)
2-phenyl-1-propene	(+)-R*-Sal	1/8	120	40	40	13	~100	11	3-phenylbutanal	~100	2.5	(S)	6)
2-phenyl-1-butene	(+)-R*-Sal	1/8	120	40	40	11.5	~100	6	3-phenylpentanal	42	1.4	(S)	15)
									2-methyl-2-phenyl-butanal	8	n.d.		
									4-phenylpentanal	50	<0.1	(R)	
1-butene	(+)-R*-Sal	1/8	120	40	40	3	n.d.	61	2-methylbutanal	25	<0.1	(S)	15)
									n-pentanal	75			
styrene[e]	(+)-R*-Sal	1/8	90	50	50	3.5	95	32	2-phenylpropanal diethyl acetal	75	15.2	(R)	15)
									3-phenylpropanal diethyl acetal	25			
2-phenyl-1-propene[e]	(−)-DIOP	1/8	90	40	40	3.5	~100	5	3-phenylbutanal diethyl acetal	~100	0.6	(R)	15)
(Z)-2-butene[f]	(−)-DIOP	1/1	150	50	2	30	n.r.	12	2-methylbutanal	40	2.7	(R)	16)
									n-pentanal	60			
bicyclo[2,2,2]-oct-2-ene	(−)-DIOP	1/1	150	50	2	10	n.r.	35	bicyclo[2,2,2]octane-2-carbaldehyde	~100	1.2	(R)	16)

[a] $[Co(CO)_4]_2$ as the catalyst precursor unless stated otherwise.

[b] $\dfrac{\text{Moles reacted olefin}}{\text{Moles starting olefin}} \times 100$.

[c] $\dfrac{\text{Moles aldehydes}}{\text{Moles reacted olefin}} \times 100$.

[d] $\dfrac{\text{Moles aldehyde considered}}{\text{Moles aldehydic products}} \times 100$.

[e] Ethanol as the solvent in the presence of ethyl orthoformate.

[f] $HCo(CO)_4$ as catalyst precursor. n.d. = not determined; n.r. = not reported

changing the solvent from a hydrocarbon to ethanol have not been explained so far [19].

The fact that 2-phenylbutane, which is the main product formed during 2-phenyl-1-butene hydroformylation, is optically inactive while the aldehydes arising from the same reaction are optically active, has been taken as an indication that the intermediate in hydrogenation is either a configurationally labile α,α-disubstituted benzyl cobalt complex [15] or the corresponding free radical [20].

2.1.2 Rhodium-Catalyzed Hydroformylation

The greater versatility of rhodium catalysts [21] and the promising results obtained in the first experiments in asymmetric hydroformylation [5, 22–24] encouraged research in this field. A great number of substrates have been hydroformylated and a relatively large number of chiral ligands have been used. Furthermore, the influence of the reaction conditions on the optical yield has been investigated for some substrates.

In Table 2 the highest optical yields obtained up to now for different substrates are reported. Relatively good optical yields have been obtained for terminal and for some internal olefins. Similar optical yields are also reported in the hydroformylation of conjugated diolefins, but in this case asymmetric induction seems to occur in the hydrogenation of the non-chiral unsaturated aldehyde which is the primary reaction product [27].

A rather large number of asymmetric ligands have been tested in the hydroformylation of styrene (Table 3), optical yields between 20 and 30% being easily obtained.

The influence of temperature and carbon monoxide partial pressure on the optical purity of the products is different, depending on the type of ligand used and on the molar ratio between ligand and metal. In the hydroformylation of styrene by [Rh(CO)$_2$Cl]$_2$ and (R)-BzMePhP* the effect of the p_{CO} was studied [34] at 110 °C and at 50 atm p_{H_2}. For ratios L*/M of 2 and 3, the optical yield shows a maximum at about 70 atm p_{CO} in the range 0–100 atm, and at higher L*/M ratio, the maximum appears at higher pressures; for a L*/M ratio of 4 there is practically no difference in the optical yield observed in experiments carried out at p_{CO} of 30 and 70 atm [34].

When (—)-DIOP is used as the chiral ligand in the hydroformylation of 1-octene under 40 atm p_{H_2} and 60 °C, the optical yield decreases from 5.9 to 4.8% when p_{CO} is increased from 5 to 120 atm. Furthermore, in the presence of (—)-DIOP, a rise in p_{H_2} brings about an increase of the optical yield [35, 41] (6.2% compared to 8.3% in the hydroformylation of 1-octene under 5 atm p_{CO} and 40 °C when the p_{H_2} is raised from 1 atm to 60 atm).

An increase of the temperature causes a decrease of the extent of the asymmetric induction in the hydroformylation of styrene when either DIOP [35, 41] or BzMePhP [34] is used as the chiral ligand. With NMDPP [34] or CHDBP [39] a rise in the temperature causes a change of the sign of the asymmetric induction. Interestingly, varying the ligand-to-metal ratio results in a change of the sign of the asymmetric induction in the presence of NMDPP whereas for (+)-BzMePhP, changing the L/M ratio from 1 to about 6 causes an increase of the optical purity of hydratropaldehyde produced. In the latter case, the optical activity remains practically unaffected for L*/M ratios greater than 6 [34]. Hydrogen and carbon monoxide pressure affects the

Table 2. Highest optical yields obtained in the rhodium-catalyzed asymmetric hydroformylation

Substrate	Catalyst precursor	Chiral ligand	Reaction conditions			
			t (°C)	p_{H_2} (atm)	p_{CO} (atm)	time (hrs)
1-butene	$[Rh(CO)_2Cl]_2$	(−)-DIOP-DBP[d]	80	50	50	10
1-pentene	$RhH(CO)(PPh_3)_3$	(−)-DIOP[e]	25	0.5	0.5	100
3-methyl-1-butene	$RhH(CO)(PPh_3)_3$	(−)-DIOP[e]	25	0.5	0.5	47
1-octene	$RhH(CO)(PPh_3)_3$	(−)-DIOP[e]	25	0.5	0.5	120
2-methyl-1-butene	$RhH(CO)(PPh_3)_3$	(−)-DIOP[e]	95	45	45	n.r.
2,3-dimethyl-1-butene	$[Rh(NBD)Cl]_2$	(−)-CHIRAPHOS[d]	100	40	40	168
2-ethyl-1-hexene	$RhH(CO)(PPh_3)_3$	(−)-DIOP[e]	100	40	40	70
vinyl acetate	$[Rh(CO)_2Cl]_2$	(−)-DIOP[d]	70	50	50	42
metallyl alcohol	$RhH(CO)(PPh_3)_3$	(−)-DIOP[e]	40	0.5	0.5	31
(Z)-2-butene	$RhH(CO)(PPh_3)_3$	(−)-DIOP[e]	25	0.33	0.33	720
(E)-2-butene	$[Rh(NBD)Cl]_2$	(−)-CHIRAPHOS[d]	100	40	40	97
(Z)-2-pentene	$[Rh(COD)((−)\text{-}DIOP)]\,BF_4$[g]		80	4	4	n.r.
(Z)-2-hexene	$RhH(CO)(PPh_3)_3$	(−)-DIOP[e]	95	41	41	95
(E)-2-hexene	$RhH(CO)(PPh_3)_3$	(−)-DIOP[e]	95	41	41	123
bicyclo[2,2,2]-oct-2-ene	$RhH(CO)(PPh_3)_3$	(−)-DIOP[e]	95	40	40	24
2,5-dihydrofurane	$RhH(CO)(PPh_3)_3$	(−)-DIOP[e]	25	0.5	0.5	120
styrene	$[Rh(1,5\text{-}HD)Cl]_2$	(+)-BzMePhP[e]	90	70	70	8
allylbenzene	$RhH(CO)(PPh_3)_3$	(−)-DIOP[e]	60	0.5	0.5	193
2-phenyl-1-propene	$[Rh(NBD)Cl]_2$	(−)-CHIRAPHOS[d]	100	40	40	70
2-phenyl-1-butene	$RhH(CO)(PPh_3)_3$	(−)-DIOP[e]	60	0.5	0.5	894
(E)-1-phenyl-1-propene	$RhH(CO)(PPh_3)_3$	(−)-DIOP[e]	60	0.5	0.5	522
phenyl vinylether	$Rh(CO)Cl(NMDPP)_2$[d]		90	70	70	n.r.
indene	$[Rh(CO)_2Cl]_2$	(−)-MePr^{n}PhP[h]	80	100	100	n.r.
isoeugenol	$[Rh(CO)_2Cl]_2$	(+)-MePr^{n}PhP[g]	80	100	100	n.r.
cinnamic alcohol	$[Rh(CO)_2Cl]_2$	(+)-MePr^{n}PhP[l]	80	100	100	n.r.

of different substrates

Con-ver-sion[a] %	Yield[b] %	Hydroformylation products and isomeric composition[c]		Chiral reaction product o.p. %	abs. conf.	Ref.
67	n.r.	2-methylbutanal n-pentanal	32 68	20.4	S	25)
n.d.	53	2-methylpentanal n-hexanal	7 93	19.7	R	26)
n.d.	25	2,3-dimethylbutanal 4-methylpentanal	7 93	15.2	R	26)
n.d.	35	2-methyloctanal n-nonanal	10 90	16.5	R	26)
	40	3-methylpentanal	n.r.	0.2	R	27)
n.d.	43	3,4-dimethylpentanal	~100	21.8	R	28)
n.d.	80	3-ethylheptanal	~100	1.1	R	15)
n.r.	13	2-acetoxypropanal 3-acetoxypropanal	92 8	23	S	29)
13	12	3-hydroxymethylbutanal	~100	14.2	S	30)
n.d.	1.7	2-methylbutanal	~100	27.0[f]	S	26)
n.d.	40	2-methylbutanal	~100	18.5	S	28)
n.r.	63	2-methylpentanal 2-ethylbutanal	70 30	> 0.1	R	32)
70	66	2-methylhexanal 2-ethylpentanal	60 40	7.6 5.8	S R	26)
50	47	2-methylhexanal 2-ethylpentanal	58 42	1.4 2.9	S R	26)
98	95	bicyclo[2,2,2]octene carbaldehyde	~100	4.2	R	33)
20	12	3-tetrahydrofurfural	~100	7.0	R	33)
n.r.	26.7	2-phenylpropanal 3-phenylpropanal	89 11	28.3	S	34)
68	65	2-methyl-3-phenylpropanal 4-phenylbutanal	12 88	15.5	R	24)
30	30	3-phenylbutanal	~100	21.4	R	28)
60	57	3-phenylpentanal	96 <	1.8	R	24)
70	68	2-methyl-3-phenylpropanal 2-phenylbutanal	12 88	14.4	R	24)
n.r.	n.r.	2-phenoxypropanal	57	0.3	R	22)
n.r.	n.r.	1-formylindane 2-formylindane	95 5	n.d.	(+)[i]	5)
n.r.	n.r.	2[2H3MP]-butanal[j] 2[2H3MB]-propanal[k]	85 15	n.d.	(−)[i]	5)
n.r.	n.r.	2-phenyl-4-hydroxypropanal	n.r.	n.d.	(−)[i]	5)

Table 2 (Continued)

Substrate	Catalyst precursor	Chiral ligand	Reaction conditions			
			t (°C)	p_{H_2} (atm)	p_{CO} (atm)	time (hrs)
2-[2-phenylvinyl]-5-methyl-1,3-dioxolane	[Rh(CO)$_2$Cl]$_2$	(+)-MePr^n PhP^l	80	100	100	n.r.
1,3-butadiene	RhH(CO) (PPh$_3$)$_3$	(—)-DIOP^d	95	45	45	60
isoprene	RhH(CO) (PPh$_3$)$_3$	(—)-DIOP^d	95	45	45	160
2,3-dimethyl-1,3-butadiene	RhH(CO) (PPh$_3$)$_3$	(—)-DIOP^d	95	45	45	160
norbornene	[Rh(NBD)Cl]$_2$	(—)-CHIRAPHOS^d	100	40	40	2
3-cyanopropene	RhH(CO) (PPh$_3$)$_3$	(—)-DIOP^d	40	0.5	0.5	7.5
2-butyne	RhH(CO) (PPh$_3$)$_3$	(—)-DIOP	80	50	50	24
1-octyne	RhH(CO) (PPh$_3$)$_3$	(—)-DIOP	95	40	40	24
phenylacetylene	RhH(CO) (PPh$_3$)$_3$	(—)-DIOP	95	43	43	28
N-vinylsuccinimide	RhH(CO) (PPh$_3$)$_3$	(—)-DIOP^m	48	16	16	48

n.r. = not reported; n.d. = not determined
[a] See footnote b in Table 1; [b] See footnote c in Table 1; [c] See footnote d in Table 1; [d] Metal-to-ligand molar ratio = 1/2; [e] Metal-to-ligand molar ratio = 1/4; [f] 2-Methylbutanol having 28.4% o.p. has been obtained with rhodium catalysts using a styrene-divinylbenezene copolymer containing 2-[p-phenyl-4,5-bis(diphenylphosphinomethyl)]-1,3-dioxolane groups as the chiral ligand [31]; [g] Metal-to-ligand molar ratio = 1; [h] Metal-to-ligand molar ratio = 1/1.35;

asymmetric induction in the styrene hydroformylation in the presence of (+)-MePhPr^{n}P. In the range 50–500 atm total pressure (p_{CO}/p_{H_2} = 1), there is a maximum of the optical yield at about 200 atm [8].

2.1.3 Platinum-Catalyzed Hydroformylation

The first asymmetric hydroformylation with platinum catalysts was carried out [42] using NMDPP as the asymmetric ligand. An optical yield of ~9% was obtained in the hydroformylation of 2-methyl-1-butene to 3-methylpentanal. Subsequently, different types of olefins were asymmetrically hydroformylated using a catalytic system formed from [(—)-DIOP]PtCl$_2$ and SnCl$_2$ · 2 H$_2$O "in situ" [42,45] (Table 4).

As shown by contradictory reports [45,48] on the prevailing chirality, optical yield and reaction rate with 1-butene and (Z)- and (E)-2-butene, the catalytic system used seems to give non-reproducible results due mostly to the quality of SnCl$_2$ · H$_2$O employed. A high reaction rate and reproducible results were obtained using [(—)-DIOP]Pt(SnCl$_3$) Cl [49] as catalyst precursor [43,44]. As shown in Table 4, the

Con-version[a] %	Yield[b] %	Hydroformylation products and isomeric composition		Chiral reaction product		Ref.
				o.p. %	abs. conf.	
n.r.	n.r.	2-[2-Ph-2-fm]-dioxolane[n]	55	n.d.	(—)[i]	5)
		2-[2-Ph-1-fm]-dioxolane[o]	45			
82	41	2-methylbutanal	75	0.1	S	27)
		n-pentanal	25			
52	25	3-methylpentanal	n.r.	32.3	S	27)
40	18	3,4-dimethylpentanal	n.r.	5.7	R	27)
~100	~100	exo-bicyclo[2,2,1]heptane 2-carboxaldehyde	~97	16.4	(1S, 2S, 4R)	28)
n.d.	22	3-cyano-2-methyl-propanal	45	21.5	R	35)
		4-cyanobutanal	55			
20	n.r.	2-methylbutanal	68	0.2	S	36)
		(E)-2-methyl-2-butanal	32			
78	n.r.	2-methyloctanal	26.5	0.2	S	36)
		n-nonanal	73.5			
99	n.r.	2-phenylpropanal	62	0.9	R	36)
		3-phenylpropanal	38			
27	n.r.	2-succinimidopropanal	n.r.	18.4	R	37)

[i] Sign of the optical rotation of the product; [j] 2-(2-methoxy-3-hydroxyphenyl)butanal; [k] 2-(3-hydroxy-2-methoxybenzyl)propanal; [l] Metal-to-ligand molar ratio 1/1.25; [m] Metal-to-ligand molar ratio = 1/4.2; [n] 2-[2-phenyl-2-formylethyl]-5-methyl-1,3-dioxolane; [o] 2-[2-phenyl-1-formylethyl]-5-methyl-1,3-dioxolane

chirality obtained starting with 1-butene is opposite to that with (Z)- or (E)-2-butene which confirms the data of Ogata et al. [48].

However, aldehydes with the same prevalent (S) configuration and low, very similar optical purity can arise from the same substrates at 100 °C and at very high conversion, thus confirming the results of a preliminary report [45].

The optical yield obtained starting with 1-butene at low conversion is the highest ever reported (~47% at 60 °C) for asymmetric hydroformylation of an aliphatic olefin, the (R) antipode of 2-methylbutanal predominating.

The low optical yield and the prevalence of the (S) antipode in the product obtained from the same substrate at very high conversion can be explained by the formation of small quantities of 2-methylbutanal with prevailing (R) configuration and relatively high optical purity together with a larger amount of the same aldehyde with prevailing (S) configuration and lower optical purity from the 2-butenes arising from isomerization of the original substrate.

The influence of the structure of the ligand [48] and of p_{H_2} and p_{CO} on the optical yield and on the selectivity of the reaction with respect to the formation of aldehydes

Table 3. Asymmetric hadroformylation of styrene by rhodium catalysts

Chiral ligand	Catalyst precursor	Reaction conditions			Isomeric composition[k]	Hydratropaldehyde		Ref.
		t (°C)	p_{H_2} (atm)	p_{CO} (atm)		o.p.[a] %	abs. conf.	
(−)-CHIRAPHOS	[Rh(NBD)Cl]$_2$	100	40	40	6/94	24.2	(R)	28)
(−)-CBDPP	Rh$_4$(CO)$_{12}$	80	50	50	29/71	2.8	(R)	39)
CBDBP[b]	Rh$_4$(CO)$_{12}$	80	50	50	14/86	27.1	(S)	39)
(−)-DIOP	[Rh(CO)$_2$Cl]$_2$	80	50	50	35/65	18.3	(R)	39)
(−)-DIOP-DBP	[Rh(CO)$_2$Cl]$_2$	80	50	50	10/90	27.6	(S)	39)
(+)-CHDPP	[Rh(CO)$_2$Cl]$_2$	90	50	50	17/93	10.2	(S)	39)
CHDBP[c]	[Rh(CO)$_2$Cl]$_2$	80	50	50	9/91	0.4	(R)	39)
(+)-CHDPPO	[Rh(CO)$_2$Cl]$_2$	80	50	50	n.r.	0.5	(S)	39)
(−)-MePhPr^{n}P	[Rh(CO)$_2$Cl]$_2$	80	100	100	n.r.	21.0	(S)	5)
M((S) 2MB) PP^d	[Rh(CO)$_2$Cl]$_2$	80	100	100	n.r.	0.3	(R)	5)
(+)-NMDPP	Rh(CO)(NMDPP)$_2$Cl	75	50	50	3/97	1.9	(R)	22)
(+)-BzMePhP	[Rh(1,5-HD)Cl]$_2$	60	70	70	2/98	28.3	(S)	34)
PAMP[e]	Rh$_4$(CO)$_{12}$	100	16	16	10/90	0.5	(S)	32)
CAMP[e]	[Rh(COD)(CAMP)$_2$]BF$_4$	100	4	4	25/75	12.3	(S)	32)
Ps-DIOP[f]	RhH(CO)(PPh$_3$)$_3$	50	30	30	5/95	2.0	(R)	40)
Pi-DIPHOL[g]	[Rh(CO)$_2$Cl]$_2$	80	50	50	11/89	27.7	(R)	31)

[a] The values of the optical purity are in many cases different from those reported in the original papers since they have been recalculated assuming in all cases $\alpha_D^{25} = 238°$ (1 = 1, neat) as the maximum value for the rotatory power of hydratropaldehyde [38];
[b] (R,R) enantiomer;   [c] (S,S) enantiomer;   [d] The absolute configuration at the P atom is unknown;   [e] Chirality not specified;
[f] Soluble polystyrene containing (4R, 5R)-2-[p-phenyl-4,5-bis(diphenylphosphinomethyl)]-1,3-dioxolane groups;
[g] Insoluble styrene-divinylbenzene copolymer containing (4R, 5R)-2-[p-phenyl-4,5-bis(5-H-dibenzophosphol-5-ylmethyl)]-1,3-dioxolane groups (J. K. Stille, personal communication);
[k] $\frac{\text{Straight chain aldehyde}}{\text{Branched aldehyde}} \times 100$

[46,47)] was investigated by using the $L_2PtCl_2/SnCl_2 \cdot 2 H_2O$ catalytic system prepared "in situ" (L_2 = DIOP or DIOP-DBP).

Changing the ligand structure by substituting two dibenzophosphole groups for two diphenylphosphinyl groups results in a variation of the prevailing chirality in butenes but not in styrene [48).

Furthermore, the above change in the structure of the ligand causes in the case of 1-butene a great increase (about 5 times) and with (Z)- and (E)-2-butene an even larger decrease of the optical yield. Part of this effect might be connected with the different extents of isomerization of the substrates in the two cases; in fact, with DIOP-DBP, a three times longer reaction time is required to reach the same conversion of olefin to aldehyde [48).

The influence of p_{CO} and p_{H_2} on the selectivity and optical yield of the reaction were investigated in the case of 2-phenyl-1-propene and 2-phenyl-1-butene [46,47).

The extent of hydroformylation versus hydrogenation rises with increasing p_{CO} and decreasing p_{H_2}; correspondingly, the optical yield decreases with increasing p_{CO} and dropping p_{H_2} [46). No effect of excess of phosphine on the optical yield was observed. In the case of 2-phenyl-1-butene, both hydrogenation and hydroformylation products are optically active, the predominating antipodes arising from opposite enantiofaces of the substrate in the two reactions; remarkably, the optical purity of the hydrogenation product is not affected by p_{CO} and p_{H_2} [47).

2.1.4 Hydroformylation by Ruthenium, Iridium and Palladium Catalysts

Very few results have been published in the field of the asymmetric hydroformylation with catalytic systems containing the above metals.

With $H_4Ru_4(CO)_8[(-)-DIOP]_2$, very low optical yields have been obtained in the hydroformylation of bicyclo[2,2,2]oct-2-ene and of (Z)-2-butene [16). With the ligand (—)-DIOP, (R) prevailing chirality has been obtained using bicyclo[2,2,2]oct-2-ene as the substrate with Ru, Rh, and Co as catalysts. In the case of (Z)-2-butene the prevailing configuration is (R) with Ru and Co, but (S) with Rh catalysts [16).

No asymmetric synthesis was achieved with the above substrates when either (S)-2-methylbutyl-diphenylphosphine or NMDPP or 2-exo-3-exo-bis(diphenylphosphinoxy)camphane were used instead of DIOP with the above ruthenium complex [16).

Iridium catalysts have been employed only infrequently in asymmetric hydroformylation; with a catalytic system prepared "in situ" from $[Ir(CO)_3(PPh_3)_2]BPh_4$ and DIOP, a low optical yield (0.5 %) was obtained in the synthesis of 2-acetoxypropanal from vinyl acetate [32).

The asymmetric hydroformylation using Pd complexes has produced conflicting results [50). A more thorough general knowledge of the Pd-catalyzed hydroformylation is needed in order to interpret the results.

2.1.5 Enantioface Discrimination and Asymmetric Induction

In asymmetric reactions in which only one isomer is formed, enantioface discrimination corresponds to asymmetric induction. However, in reactions in which two isomers are formed if both isomers are chiral, as in the hydroformylation of (Z)-2-hexene, the optical purities of the two isomers are in general different and enantioface discrimination does not correspond quantitatively (and sometimes not even quali-

Table 4. Platinum-catalyzed asymmetric hydroformylation

Substrate	Catalyst precursor	Reaction conditions				Con-ver-sion[a]	Yield[a]	Hydroformylation products and isomeric composition[a]		Chiral reaction product		Ref.
		t (°C)	P_{H_2} (atm)	Pco (atm)	time (hrs)					o.p. %	abs. conf.	
1-butene	[(−)-DIOP]Pt(SnCl$_3$)Cl	60	40	40	6.5	40	32	2-methylbutanal n-pentanal	3.5 96.5	46.7	(R)	43)
1-butene	[(−)-DIOP]Pt(SnCl$_3$)Cl	100	40	40	1.4	90	68	2-methylbutanal n-pentanal	15 85	3.2	(S)	43)
(Z)-2-butene	[(−)-DIOP]Pt(SnCl$_3$)Cl	60	40	40	5.5	32	31	2-methylbutanal n-pentanal	76 24	14.5	(S)	44)
(Z)-2-butene	[(−)-DIOP]Pt(SnCl$_3$)Cl	100	40	40	3.3	95	89	2-methylbutanal n-pentanal	36 64	9.3	(S)	43)
(E)-2-butene	[(−)-DIOP]Pt(SnCl$_3$)Cl	60	40	40	8.0	36	35	2-methylbutanal n-pentanal	77 23	24.2	(S)	44)
(E)-2-butene	[(−)-DIOP]Pt(SnCl$_3$)Cl	100	40	40	4.7	90	82	2-methylbutanal n-pentanal	40 60	10.9	(S)	43)
1-pentene	[(−)-DIOP]Pt(SnCl$_3$)Cl	60	40	40	3.2	40	35	2-methylpentanal n-hexanal	3 97	29.1	(R)	43)
1-octene	[(−)-DIOP]Pt(SnCl$_3$)Cl	60	40	40	2	12	10	2-methyloctanal n-nonanal	4 96	18.6	(R)	43)
2-methyl--1-butene	(NMDPP)$_2$PtCl$_2$ + SnCl$_2$^{d}	n.r.	100	100	n.r.	n.r.	75	3-methylpentanal	n.r.	9.4	(R)	42)
2,3-dimethyl--1-butene	[(−)-DIOP]Pt(SnCl$_3$)Cl	80	40	40	4	n.d.	80	3,4-dimethyl-pentanal	~100	19.9	(R)	45)
styrene	[(−)-DIOP]Pt(SnCl$_3$)Cl	60	125	125	5	50	45	2-phenylpropanal 3-phenylpropanal	31 69	28.6	(S)	43)
2-phenyl--1-propene	[(−)-DIOP]PtCl$_2$ + SnCl$_2$^{b,c}	100	200	50	4	98	63	3-phenylbutanal	~100	13.2	(S)	46)
2-phenyl--1-butene	[(−)-DIOP]PtCl$_2$ + SnCl$_2$^{b,c}	100	200	50	4	~100	48	3-phenylpentanal	~100	20.7	(S)	47)
norbornene	[(−)-DIOP]Pt(SnCl$_3$)Cl	80	40	40	3.5	40	22					

Table 4. Platinum-catalyzed asymmetric hydroformylation

Substrate	Catalyst precursor	Reaction conditions				Con-version[a]	Yield[a]	Hydroformylation products and isomeric composition[a]		Chiral reaction product		Ref.
		t (°C)	P_{H_2} (atm)	Pco (atm)	time (hrs)					o.p. %	abs. conf.	
1-butene	[(−)-DIOP]Pt(SnCl$_3$)Cl	60	40	40	6.5	40	32	2-methylbutanal	3.5	46.7	(R)	43)
								n-pentanal	96.5			
1-butene	[(−)-DIOP]Pt(SnCl$_3$)Cl	100	40	40	1.4	90	68	2-methylbutanal	15	3.2	(S)	43)
								n-pentanal	85			
(Z)-2-butene	[(−)-DIOP]Pt(SnCl$_3$)Cl	60	40	40	5.5	32	31	2-methylbutanal	76	14.5	(S)	44)
								n-pentanal	24			
(Z)-2-butene	[(−)-DIOP]Pt(SnCl$_3$)Cl	100	40	40	3.3	95	89	2-methylbutanal	36	9.3	(S)	43)
								n-pentanal	64			
(E)-2-butene	[(−)-DIOP]Pt(SnCl$_3$)Cl	60	40	40	8.0	36	35	2-methylbutanal	77	24.2	(S)	44)
								n-pentanal	23			
(E)-2-butene	[(−)-DIOP]Pt(SnCl$_3$)Cl	100	40	40	4.7	90	82	2-methylbutanal	40	10.9	(S)	43)
								n-pentanal	60			
1-pentene	[(−)-DIOP]Pt(SnCl$_3$)Cl	60	40	40	3.2	40	35	2-methylpentanal	3	29.1	(R)	43)
								n-hexanal	97			
1-octene	[(−)-DIOP]Pt(SnCl$_3$)Cl	60	40	40	2	12	10	2-methyloctanal	4	18.6	(R)	43)
								n-nonanal	96			
2-methyl-1-butene	(NMDPP)$_2$PtCl$_2$ + SnCl$_2$[d]	n.r.	100	100	n.r.	n.r.	75	3-methylpentanal	n.r.	9.4	(R)	42)
2,3-dimethyl-1-butene	[(−)-DIOP]Pt(SnCl$_3$)Cl	80	40	40	4	n.d.	80	3,4-dimethyl-pentanal	~100	19.9	(R)	45)
styrene	[(−)-DIOP]Pt(SnCl$_3$)Cl	60	125	125	5	50	45	2-phenylpropanal	31	28.6	(S)	43)
								3-phenylpropanal	69			
2-phenyl-1-propene	[(−)-DIOP]PtCl$_2$ + SnCl$_2$[b,c]	100	200	50	4	98	63	3-phenylbutanal	~100	13.2	(S)	46)
2-phenyl-1-butene	[(−)-DIOP]PtCl$_2$ + SnCl$_2$[b,c]	100	200	50	4	~100	48	3-phenylpentanal	~100	20.7	(S)	47)
norbornene	[(−)-DIOP]Pt(SnCl$_3$)Cl	80	40	40	3.5	40	33	exo-bicyclo[2,2,1]-heptane-2-carbaldehyde	~98	28.4	(1R, 2R, 4S)	43)

[a] See corresponding notes in Table 2;   [b] Molar ratio L$_2$PtCl$_2$/SnCl$_2$ · 2 H$_2$O = 1:5;   [c] SnCl$_2$ · 2 H$_2$O has been used

[46,47)] was investigated by using the $L_2PtCl_2/SnCl_2 \cdot 2 H_2O$ catalytic system prepared "in situ" ($L_2 = $ DIOP or DIOP-DBP).

Changing the ligand structure by substituting two dibenzophosphole groups for two diphenylphosphinyl groups results in a variation of the prevailing chirality in butenes but not in styrene [48].

Furthermore, the above change in the structure of the ligand causes in the case of 1-butene a great increase (about 5 times) and with (Z)- and (E)-2-butene an even larger decrease of the optical yield. Part of this effect might be connected with the different extents of isomerization of the substrates in the two cases; in fact, with DIOP-DBP, a three times longer reaction time is required to reach the same conversion of olefin to aldehyde [48].

The influence of p_{CO} and p_{H_2} on the selectivity and optical yield of the reaction were investigated in the case of 2-phenyl-1-propene and 2-phenyl-1-butene [46,47].

The extent of hydroformylation versus hydrogenation rises with increasing p_{CO} and decreasing p_{H_2}; correspondingly, the optical yield decreases with increasing p_{CO} and dropping p_{H_2} [46]. No effect of excess of phosphine on the optical yield was observed. In the case of 2-phenyl-1-butene, both hydrogenation and hydroformylation products are optically active, the predominating antipodes arising from opposite enantiofaces of the substrate in the two reactions; remarkably, the optical purity of the hydrogenation product is not affected by p_{CO} and p_{H_2} [47].

2.1.4 Hydroformylation by Ruthenium, Iridium and Palladium Catalysts

Very few results have been published in the field of the asymmetric hydroformylation with catalytic systems containing the above metals.

With $H_4Ru_4(CO)_8[(-)-DIOP]_2$, very low optical yields have been obtained in the hydroformylation of bicyclo[2,2,2]oct-2-ene and of (Z)-2-butene [16]. With the ligand (−)-DIOP, (R) prevailing chirality has been obtained using bicyclo[2,2,2]oct-2-ene as the substrate with Ru, Rh, and Co as catalysts. In the case of (Z)-2-butene the prevailing configuration is (R) with Ru and Co, but (S) with Rh catalysts [16].

No asymmetric synthesis was achieved with the above substrates when either (S)-2-methylbutyl-diphenylphosphine or NMDPP or 2-exo-3-exo-bis(diphenylphosphinoxy)camphane were used instead of DIOP with the above ruthenium complex [16].

Iridium catalysts have been employed only infrequently in asymmetric hydroformylation; with a catalytic system prepared "in situ" from $[Ir(CO)_3(PPh_3)_2]BPh_4$ and DIOP, a low optical yield (0.5%) was obtained in the synthesis of 2-acetoxypropanal from vinyl acetate [32].

The asymmetric hydroformylation using Pd complexes has produced conflicting results [50]. A more thorough general knowledge of the Pd-catalyzed hydroformylation is needed in order to interpret the results.

2.1.5 Enantioface Discrimination and Asymmetric Induction

In asymmetric reactions in which only one isomer is formed, enantioface discrimination corresponds to asymmetric induction. However, in reactions in which two isomers are formed if both isomers are chiral, as in the hydroformylation of (Z)-2-hexene, the optical purities of the two isomers are in general different and enantioface discrimination does not correspond quantitatively (and sometimes not even qualitatively) to the enentiomeric excess measured for the two isomers [12] (Fig. 1). This fact is readily understood if we take into account that the 2 pairs of antipodes can form through 4 independent transition states; the activation energy determining the rate constant for the formation of each of the 4 reaction products is usually different and therefore no relationship must necessarily exist between the concentrations of the products. If the reaction occurs stepwise, e.g. via the preliminary formation of π-olefin complexes of the metal of the catalytic complex, as is generally supposed to be the case in hydroformylation [21], the situation can be suitably described considering differences in regioselectivity of the reaction occurring on either enantioface of the substrate [12]. On this basis, when only one of the regioisomeric hydroformylation products is chiral, the enantiomeric excess measured is

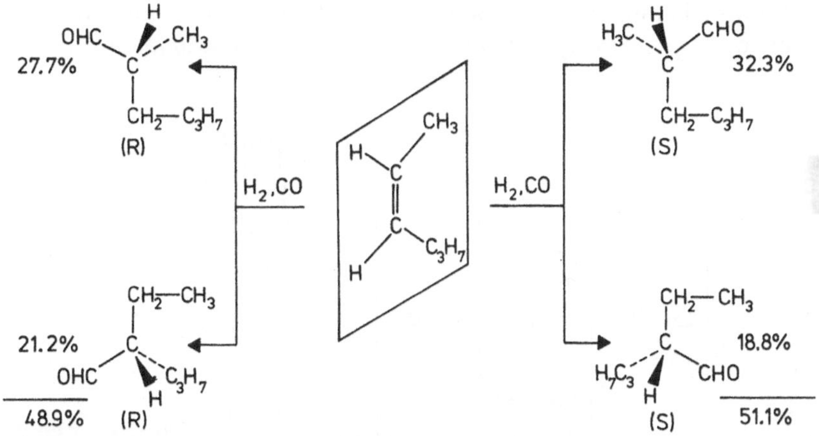

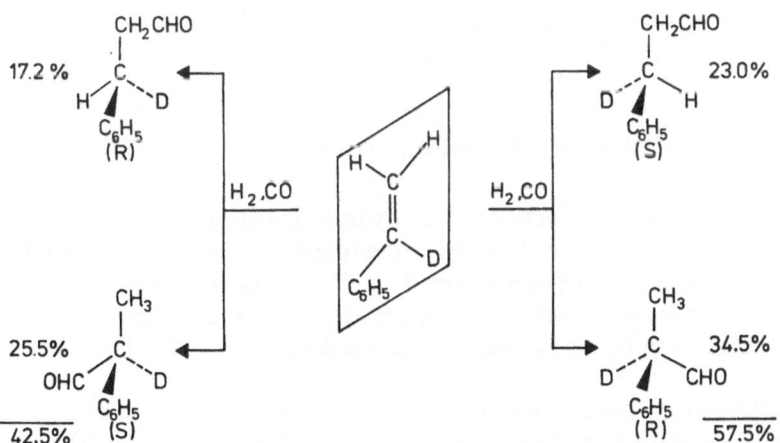

Fig. 1. Regio- and stereoisomeric composition of the aldehydes arising from asymmetric hydroformylation of (Z)-2-hexene catalyzed by Rh/(—)-DIOP [15]

Fig. 2. Regio- and stereoisomeric composition of the aldehydes arising from asymmetric hydroformylation of α-[²H]-styrene catalyzed by Rh/(—)-DIOP [51]

91

not necessarily related qualitatively and quantitatively to the enantioface discrimination. In this case, in order to determine the extent of enantioface discrimination, a chirality center must be created also in the linear product either by using labelled olefins or by carrying out a deuterioformylation instead of hydroformylation.

For instance, when α-[²H]-styrene is hydroformylated with the Rh/(—)-DIOP catalytic system, the two reaction products obtained have almost identical optical purity and opposite absolute configuration (Fig. 2) [51]. Therefore, in this case, the enantiomeric excess measured indicates both the type (the re-re enantioface reacts preferentially) and extent of enantioface discrimination (∼15%) occurring during the reaction.

In the case of the deuterioformylation of 1-pentene in the presence of [(—)-DIOP] Pt(SnCl₃) Cl as the catalyst precursor, on the other hand, the two chiral reaction products (Fig. 3) mainly arise from the two opposite enantiofaces [43]. Therefore, in 1-pentene hydroformylation, the enantiomeric excess found in the chiral reaction product does not correspond in type and extent to enantioface discrimination.

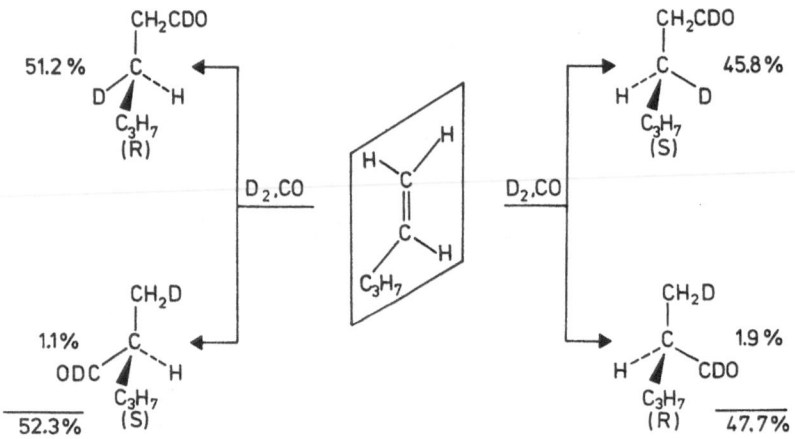

Fig. 3. Regio- and stereoisomeric composition of the aldehydes arising from asymmetric deuterioformylation of 1-pentene catalyzed by [(—)-DIOP]Pt(SnCl₃)Cl [43]

2.2 Enantiomer-Discriminating Hydroformylation

The purpose of most of the investigations of the hydroformylation of racemic olefins has been to contribute to the understanding of the origin of asymmetric induction; its significance for synthetic purposes is small. In fact, with few exceptions, this type of reaction never leads to even a fair optical purity of the reaction products (stereoelective synthesis) or of the recovered substrate (kinetic resolution) (Scheme 1, reaction 3).

When more than one isomer is formed during the reaction an enantiomer discrimination can take place also with quantitative transformation of the substrate (Scheme 1, reaction 4). This is the case in the hydroformylation of 3-phenyl-1-butene with a Rh/(—)-DIOP catalytic system [14] where 4-phenylpentanal arises preferentially

Table 5. Enantiomer-differentiating reactions using (—)-DIOP as the chiral ligand

Substrate	Hydrogenation[a]			Hydroformylation[b]			Hydroformylation[c]		
	Conversion[d]	Unreacted substrate		Conversion[d]	Unreacted substrate		Conversion[d]	Unreacted substrate	
		prevail. chiral.	o.p. %		prevail. chiral.	o.p. %		prevail. chiral.	o.p. %
$C_2H_5-CH-CH=CH_2$ $\mid$ CH_3	50	(R)	<0.1	51	(S)	3.3	63	—	~0
$(n\text{-}C_3H_7)-CH-CH=CH_2$ $\mid$ CH_3	70	(R)	0.6	50	(S)	2.8			
$(i\text{-}C_3H_7)-CH-CH=CH_2$ $\mid$ CH_3				52	(S)	2.6			
$(t\text{-}C_4H_9)-CH-CH=CH_2$ $\mid$ CH_3	64	(S)	≪0.1	49	(R)	14.0			
$C_6H_5-CH-CH=CH_2$ $\mid$ CH_3	54	(S)	0.4	47	(R)	4.5			
$C_2H_5-CH-C=CH_2$ $\mid \quad \mid$ $CH_3 \ CH_3$	46	(S)	2.2	50[e]	—	0	50	(R)	0.6
$C_6H_5-CH-C=CH_2$ $\mid \quad \mid$ $CH_3 \ CH_3$	48	(R)	1.0						

[a] [Rh(NBD)Cl]$_2$ as the catalyst precursor; at 25 °C and 1 atm H$_2$;   [b] RhH(CO)(PPh$_3$)$_3$ as the catalyst precursor; at 40 °C and 1 atm H$_2$ + CO (1:1);
[c] [(—)-DIOP]PtCl$_2$ + SnCl$_2$ (1:5) as the catalyst precursor; at 100 °C and 80 atm H$_2$ + CO (1:1);   [e] At 80 °C under 80 atm H$_2$ + CO (1:1)
[d] Moles reacted olefin / Moles starting olefin × 100;

from the (S) antipode and 2-methyl-3-phenylbutanal preferentially from the (R) antipode, with almost quantitative conversion of the substrate to aldehydes (Scheme 2).

Kinetic resolutions during hydroformylation were achieved using $Co_2(CO)_8$ and (+)-R*-Sal [6], RhH(CO) (PPh$_3$)$_3$ and (—)-DIOP [15] or [(—)-DIOP]PtCl$_2$ in the presence of SnCl$_2$. The optical purity of the olefins recovered is, in general, very low but in most cases it is sufficient to determine which antipode is the one that reacts preferentially. In Table 5 the results of enantiomer discriminating hydroformylation are compared with those obtained with the same substrates in enantiomer-discriminating hydrogenation.

3 Regularities Observed in Asymmetric Hydroformylation

The large amount of data on the prevailing chirality of the reaction products obtained with the same catalytic system allows some considerations to be made on the influence of the structure of the substrate, of the type of the catalytically active metal and of the type of ligand on the type of asymmetric induction.[3]

These three aspects will be considered separately. In the case of enantioface-discriminating reactions to avoid discrepancies between prevailing chirality of the products and enantioface prevailingly reacting (see Sect. 2.1.5) the face of the prochiral atom preferentially involved [85] in the reaction has been chosen as a parameter to establish possible regularities. In the case of enantiomer-discriminating reactions attempts to determine regularities have been made on the basis of the chirality of the non-reacted substrate.

3.1 Hydroformylation of Different Substrates with the Same Catalytic System

3.1.1 Monosubstituted Ethylenes

In Table 6 the results concerning the asymmetric hydroformylation of 1-butene and of styrene with different catalytic systems are reported. When rhodium-containing catalytic systems are used in the presence of several diphosphine ligands, the face of the prochiral unsaturated carbon atom which is preferentially formylated is the same in both substrates for each chiral ligand.

The same is also true for the Co/(+)-R*-Sal and for the [(—)-DIOP-DBP]PtCl$_2$/SnCl$_2$ [48] catalytic systems. For [(—)-DIOP]Pt(SnCl$_3$) Cl [43] and [(—)-CHIRAPHOS] Pt(SnCl$_3$) Cl [52], however, opposite faces of the unsaturated carbon atom are mainly formylated in the two substrates.

Furthermore, in the case of the asymmetric catalytic system containing rhodium and (—)-DIOP always the same prochiral face (re) is preferentially formylated for six other monosubstituted olefins (Table 7, column 1). Similar results are obtained with rhodium catalysts when monophosphines are used instead of DIOP. The only

---

3 As type of asymmetric induction we mean the prevalence of a given antipode over the other

Table 6. Face of the unsaturated prochiral carbon atom preferably attacked by CO in 1-butene and styrene hydroformylation

Catalyst precursor	Chiral ligand abs. conf.	Face preferably formylated		Ref.
		1-butene	styrene	
$RhH(CO)(PPh_3)_3$ + DIOP	(R,R)	re	re	[15]
$[Rh(NBD)Cl]_2$ + CHIRAPHOS	(S,S)	re	re	[28]
$[Rh(CO)_2Cl]_2$ + CHDPPO	(S,S)	si	si	[39]
$[Rh(CO)_2Cl]_2$ + CHDPM	(S,S)	si	si	[39]
$[Rh(CO)_2Cl]_2$ + CHDBP	(S,S)	si	si[a]	[39]
$[Rh(CO)_2Cl]_2$ + DIOP-DBP	(R,R)	si	si	[39]
$Rh_4(CO)_{12}$ + CBDPM	(R,R)	re	re	[39]
$Rh_4(CO)_{12}$ + CBDBP	(R,R)	si	si	[39]
$Co_2(CO)_8$ + R*-Sal	(S)	si	si	[6]
$[DIOP]Pt(SnCl_3)Cl$	(R,R)	re	si	[43]
$[DIOP-DBP]PtCl_2$-$SnCl_2$	(R,R)	si	si	[48]
$[CHIRAPHOS]Pt(SnCl_3)Cl$	(S,S)	si	re	[52]

[a] The re face of the prochiral atom is preferably formylated at temperatures below 100 °C

exception has been found using BzMePhP as the chiral ligand for which opposite faces of the unsaturated carbon atom are predominantly formylated in styrene and vinyl acetate. With $[(-)$-DIOP]Pt($SnCl_3$)Cl, the same face of the substituted olefinic carbon atom is preferentially attacked in the case of three linear α-olefins.

On the whole, with each catalytic system, the same face of the substituted unsaturated carbon atoms is attacked for different substrates. There are 2 exceptions out of 32 cases for rhodium- and 2 exceptions out of 8 cases for platinum-containing catalytic systems.

3.1.2 1,1-Disubstituted Ethylenes

The results obtained are listed in Table 8. For identical catalytic systems, faces of the disubstituted unsaturated carbon atom having the same geometric requirements (although named differently, Fig. 4) are preferentially hydroformylated [15] in aromatic substrates. This is also true for aliphatic olefins. However, with aromatic

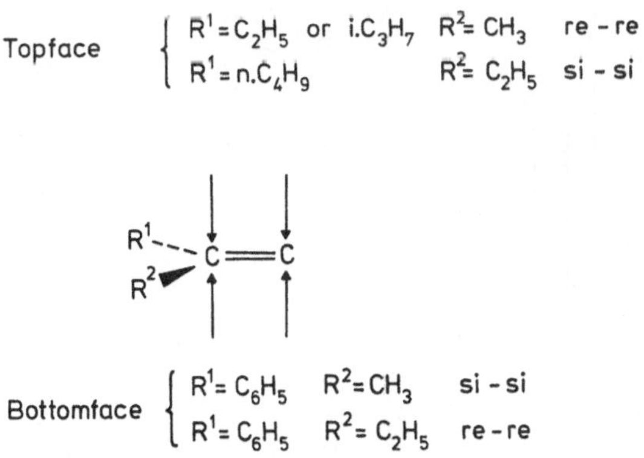

Topface
$\begin{cases} R^1 = C_2H_5 \text{ or } i.C_3H_7 \quad R^2 = CH_3 \quad re - re \\ R^1 = n.C_4H_9 \quad\quad\quad\quad R^2 = C_2H_5 \quad si - si \end{cases}$

Bottomface
$\begin{cases} R^1 = C_6H_5 \quad R^2 = CH_3 \quad si - si \\ R^1 = C_6H_5 \quad R^2 = C_2H_5 \quad re - re \end{cases}$

Fig. 4. Enantiofaces preferentially attacked in asymmetric hydroformylation of 1,1-disubstituted ethylenes with the Rh/(−)-DIOP catalytic system

Table 7. Face of the unsaturated prochiral carbon atom preferentially attacked by CO in monosubstituted ethylenes hydroformylation

Substrate	Catalytic system				
	(−)-DIOP HRh(CO)(PPh₃)₃	(+)-NMDPP [Rh(CO)₂Cl]₂	(+)-BzMePhP* [Rh(CO)₂Cl]₂	CAMP[a] [Rh(COD)L₂*]BF₄	[(−)-DIOP]PtCl₂SnCl₂
Styrene	re[15]	re[b 34] re[22]	si[34]	si[32]	si[48]
Phenyl vinyl ether					
Allylbenzene	re[15]				
1-Butene	re[15]				re[48]
3-Methyl-1-butene	re[15]				
1-Pentene	re[15]				re[43]
1-Octene	re[15]				re[43]
3-Cyanopropene	re[35]				
Vinyl acetate	re[29]	re[29]	re[29]	si[32]	
N-Vinylsuccinimide	re[37]				

a Optical rotation not reported
b si for L*/Rh ratios lower than 12

Table 8. Face of the unsaturated prochiral carbon atom which originates the predominant enantiomer in 1,1-disubstituted ethylenes hydroformylation

Substrate	Catalytic System				
	(+)-R*-SAL Co₂(CO)₈	(−)-DIOP RhH(CO)(PPh₃)₃	(−)-CHIRAPHOS [Rh(NBD)Cl]₂	[(−)-DIOP]PtCl₂ SnCl₂	[(−)-CHIRAPHOS]PtCl₂ SnCl₂
2-methyl-1-butene		re[27]	si[28]	si[43]	
2,3-dimethyl-1-butene		re[45]		si[45]	re[52]
2-ethyl-1-hexene		si[15]			
2-phenyl-1-propene	re[15]	si[15]	si[28]	re[46]	re[52]
2-phenyl-1-butene	si[15]	re[15]		si[46]	

and aliphatic olefins, opposite faces are preferentially attacked in the few examples available when (—)-DIOP is used as the chiral ligand. In all cases, with the same chiral ligand, (—)-DIOP or (—)-CHIRAPHOS, opposite faces are preferentially attacked, depending on whether rhodium- or platinum-catalysts are used.

3.1.3 Internal Olefins

The results of the hydroformylation of internal olefins are reported in Table 9. In the case of (Z)- and (E)-2-butene, the same face of the unsaturated carbon atom is formylated with either a rhodium- or platinum (—)-DIOP-containing catalytic system. With the rhodium catalyst, when an acyclic olefin is used as the substrate, the same face is always attacked, and it is only the notation but not the geometric requirement that is different for (E)-1-phenyl-1-propene. The only exception is represented by bicyclo[2,2,1]heptene. However, using (—)-CHIRAPHOS instead of (—)-DIOP, also bicyclo[2,2,1]heptene behaves like internal butenes. No regularity is observed for the cobalt or ruthenium (—)-DIOP catalytic systems. With the same system, only in 3 cases out of 15 the face of the prochiral atom preferentially formylated has different geometric requirements.

3.1.4 Racemic Olefins

With rhodium and (—)-DIOP as the chiral ligand, the chirality of the antipode preferentially reacted is opposite in hydrogenation ([Rh(NBD)Cl]$_2$ as the catalyst precursor) compared to hydroformylation (Table 5).

In hydroformylation, the optical purity of the recovered olefin is higher for monosubstituted ethylenes; in hydrogenation it is higher for 1,1-disubstituted ethylenes.

In all examples available, geometrically similar antipodes react preferentially in the case of mono- and di-substituted ethylenes respectively, the three exceptions in Table 5 being due to the fact that substrates with similar geometry have different notations (Fig. 5). No observable kinetic resolution is achieved in the platinum-catalyzed hydroformylation of 3-methyl-1-pentene whereas a slight enantiomer discrimination is observed in the case of 2,4-dimethyl-1-pentene.

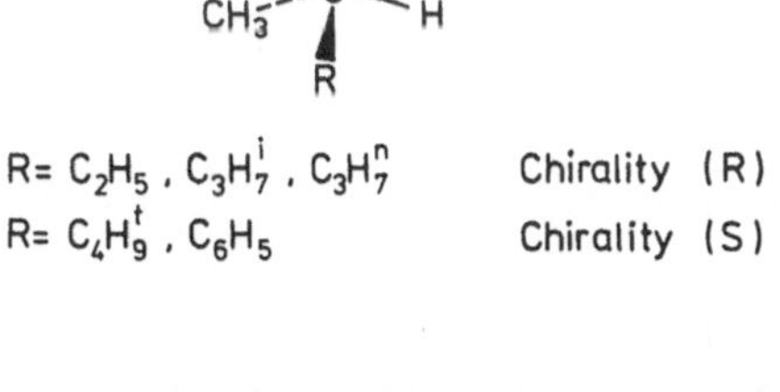

R= C_2H_5 , $C_3H_7^i$, $C_3H_7^n$ Chirality (R)

R= $C_4H_9^t$, C_6H_5 Chirality (S)

Fig. 5. Enantiomer preferentially reacting in asymmetric hydroformylation of racemic olefins using Rh/(—)-DIOP

3.2 Hydroformylation with Catalytic Systems Containing the Same Chiral Ligand and Different Metals

One of the unsolved problems in asymmetric catalysis is the origin of asymmetric induction. In fact, both an interaction between substrate and chiral ligand [53] and/or

Table 9. Face of the Unsaturated Carbon Atom Prevailingly Attacked by CO in Internal Olefins Hydroformylation

Substrate	Catalytic system					
	(−)-DIOP RhH(CO)(PPh$_3$)$_3$	(−)-CHIRAPHOS [Rh(NBD)Cl]$_2$	(−)-DIOP HCo(CO)$_4$	(−)-DIOP H$_4$Ru$_4$(CO)$_8$L$_2^*$	[(−)-DIOP]PtCl$_2$ SnCl$_2$	[(−)-DIOP-DBP]PtCl$_2$ SnCl$_2$
(Z)-2-butene	si [15]	si [28]	re [16]	re [16]	si [45]	re [48]
(E)-2-butene	si [15]	si [28]			si [45]	re [48]
(Z)-2-hexene	{ si [15] / si [15]					
(E)-2-hexene	{ si [15] / si [15]					
(E)-1-phenyl-1-propene	re [15]					
2,5-dihydrofurane	re [33]					
bicyclo-[2.2.1]-heptene	re [33]	si [28]	si [16]			
bicyclo-[2.2.2]-octene	si [33]			si [16]		

between substrate and a chiral catalytically active metal atom [54], in the step in which the asymmetric induction occurs, could be invoked to explain asymmetric induction. This subject has recently been discussed [55] and we shall limit our remarks to the insights given by asymmetric hydroformylation in the presence of platinum or rhodium catalytic systems.

As already mentioned, in the cases in which a very large excess of a single chiral product is formed, the enantiomeric excess gives a good indication of the enantioface which reacts preferentially. If we take this into consideration, it appears from the results of the asymmetric hydroformylation of 1,1-disubstituted ethylenes that opposite enantiofaces can preferentially react when Pt or Rh is used together with the same chiral ligand ((−)-DIOP or (−)-CHIRAPHOS) [52] (Table 8). The same is true when styrene is used as a substrate [28,52] (Table 7).

These results show that, at least for the hydroformylation of 1,1-disubstituted ethylenes and styrene, the geometry of the complexes, including the chirality of the metal atom and possible differences in the prevailing conformation of the ligands, is relevant in determining the asymmetric induction.

On the contrary, no influence of the metal atoms is apparent when (Z)- or (E)-2-butene are used as substrate with Rh/(−)-DIOP and Pt/(−)-DIOP catalytic systems. An attempt to rationalize the above results is presented in Section 4.

3.3 Rhodium-Catalyzed Hydroformylation of Styrene with Different Asymmetric Ligands

Styrene has been hydroformylated using a rhodium catalyst in the presence of a great number of chiral ligands. Whereas optically active β-keto enolates [56] or Shiff's bases [35] seem to be ineffective in promoting asymmetric induction, the use of phosphines or phosphinites gives rise to optically active hydratropaldehyde (Table 3). Comparing the results obtained with diphosphines having similar geometric characteristics but different descriptors (Table 3), it appears that a correlation exists between the geometry of the phosphines and the prevailing chirality of the products. In fact (−)-CHIRAPHOS is, from a geometric point of view, similar to (−)-DIP and (R) hydratropaldehyde prevails in both cases.

However, the prevailing chirality of the products in the case of the dibenzophospholes ligands is opposite to that observed with diphenylphosphines having the similar structure and the same chirality. This indicates that the chirality of the ligand is not the only factor determining the type of the prevailing chirality in the products.

With monophosphines, the results are difficult to compare because of the different reaction conditions and lack of indications of the chirality of the ligands used. The relationship between chirality of the phosphine (when known) and prevailing chirality in the products is the same as in the case of the diphosphines. Concerning the value of enantiomeric excess it appears that when the anisyl group in PAMP is exchanged for an n-propyl group (MePhPr^{n}P) or a benzyl group (BzMePhP), the optical purities of the recovered hydratropaldehyde rise from 0.5% to 21% or 28.3%, respectively.

4 A Model for the Diastereomeric Transition States Controlling Asymmetric Induction in Hydroformylation

The prevalence of one antipode in the chiral product obtained by hydroformylation of an olefin must be connected with the different free energy of the transition states in the step (or steps) in which asymmetric induction occurs.

It was shown [15] by asymmetric hydroformylation of linear butenes that asymmetric induction in hydroformylation occurs substantially before or during metal alkyl intermediate formation, which, according to the accepted hydroformylation mechanism [21,57,58], is the second step in the catalytic process (see also Sect. 5.1).

To correlate the results of asymmetric hydroformylation the following assumptions have been made:

i)   The activated complex leading to the intermediate metal-alkyl complex has the highest free energy.
ii) The metal atom approached by the substrate in the transition state is a chiral center.
iii) The double bond of the substrate in the transition state is approximately coplanar to the bond between metal and hydrogen [86,87].

The simplest way to represent such a transition state is to project it onto the plane perpendicular to the approach direction of the substrate. The groups bonded to the metal are conceived as spheres centered along the metal-ligand bonds. In Fig. 6, L and S represent two ligands of different size, L being larger than S. Z represent a third ligand, which is assumed to be different and larger than H. The space above the plane of the projection is divided into 4 regions defined by two mutually perpendicular planes, containing the approach line and the L-M-S and H-M-Z axes respectively (i.e. the 4 quadrants Q_1, Q_2, Q_{-1} and Q_{-2} shown in Fig. 6a). The space available to accommodate the groups bound to an unsaturated carbon atom of the substrate in which the double bond is superimposed and parallel to the MH bond (Fig. 6b), decreases in the order $Q_2 > Q_1 \gg Q_{-2} > Q_{-1}$. If the free energy differences between the transition states controlling asymmetric induction depend predominantly on the repulsive steric interactions between substrate and face of the catalyst approached by the substrate, the group(s) bound to the unsaturated carbon atoms of the substrate will prefer the quadrant in which more space is available. If the relative positions of the ligands L and S with respect to the M—H bond and the relative position of H and Z with respect to the L—M—S axis are known, the face of the prochiral olefin preferentially reacting with the catalyst and the prevailing enantiomer in the chiral reaction product can be predicted. The same model indicates also the position of the substrate in which the formyl group will be preferentially bound in the hydroformylation product.

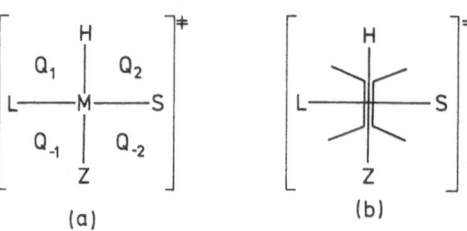

Fig. 6. Model for a transition state determining asymmetric induction in asymmetric hydroformylation

For a monosubstituted ethylene (Fig. 7) the model predicts that the linear isomer (b) must largely prevail over a + c (the branched isomer) and that antipode c must prevail over antipode a.

The relative position of L with respect to S and of H with respect to Z can be decided, for instance, by hydroformylation of (Z)-2-butene, which yields only one aldehyde that is chiral. When the catalyst is optically active, the predominating antipode in the reaction product indicates the face of the unsaturated carbon atoms preferentially attacked by CO and therefore the more stable transition state (Fig. 8) (that is, on the assumption that the difference in the free energy of the transition state mainly depends on steric interactions, the transition state in which such steric interactions are smaller).

In the hydroformylation of 1-butene (Fig. 7, R = C_2H_5) with the same catalytic system, the model gives correct qualitative information, not only about the pre-

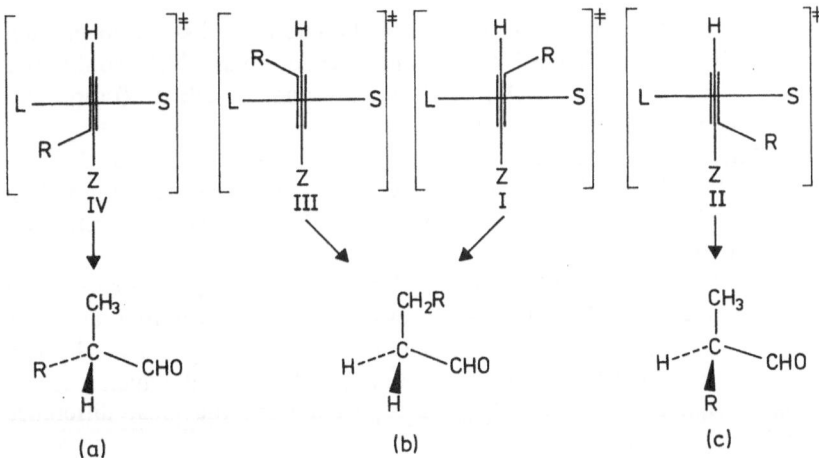

Fig. 7. Possible transition states (and corresponding products) in asymmetric hydroformylation of monosubstituted ethylenes

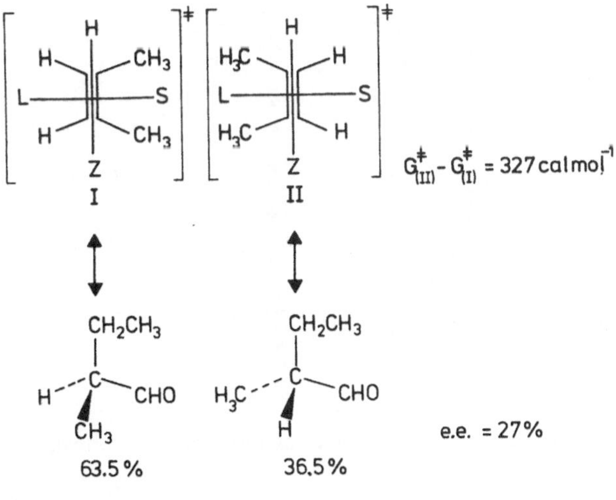

$$G_{(II)}^{\ddagger} - G_{(I)}^{\ddagger} = 327 \, cal \, mol^{-1}$$

e.e. = 27%

Fig. 8. Transition states determining asymmetric induction in asymmetric hydroformylation of (Z)-2-butene (Figures refer to the Rh/(—)-DIOP catalytic system)

dominating antipode of the chiral 2-methylbutanal formed, but also about the relative amounts of 2-methyl-butanal and n-pentanal formed. In fact, according to the model, compound b (the linear isomer) must largely prevail over the sum of compounds a and c, the two antipodes of branched isomer (experimentally found for Rh/(—)-DIOP [15] b/(a + c) = 12.5). The (S) antipode of 2-methyl-butanal (c) must prevail over a, the (R) antipode (experimentally found c/a = 1.5) [15].

4.1 Application of the Model to Different Substrates

For the application of the model to the different substrates it has been assumed that the prevailing chirality of the metal does not change by varying the substrate.

4.1.1 1,2-Disubstituted Ethylenes

From the results of the hydroformylation of (Z)-2-butene it is possible to predict, when the same catalytic system is used, the prevailing enantiomer which would arise from other C_{2v} olefins, such as bicyclo[2.2.2]oct-2-ene and 2,5-dihydrofurane and for other internal olefins like (Z)-2-hexene.

In the case of internal C_s olefins, for which 4 transition states must exist (Fig. 9), the model also predicts the carbon atom preferentially formylated (corresponding to the unsaturated carbon atom nearer to the metal atom than to the hydrogen atom of the catalyst in the transition state).

In fact, for (Z)-2-hexene, since the space available in Q_1 and Q_2 is greater than that in Q_{-1} and Q_{-2}, 2-methylhexanal must prevail over 2-ethylpentanal, as it has indeed been found (2-methylhexanal/2-ethylpentanal = 1.5). In this case, the isomeric ratio reflects the value $|(G^{\ddagger}_{(III)} + G^{\ddagger}_{(IV)}) - (G^{\ddagger}_{(I)} + G^{\ddagger}_{(II)})|$ (Fig. 9) which is larger than both $(G^{\ddagger}_{(II)} - G^{\ddagger}_{(I)})$ and $(G^{\ddagger}_{(IV)} - G^{\ddagger}_{(III)})$ as expected from the great difference in size between H and Z.

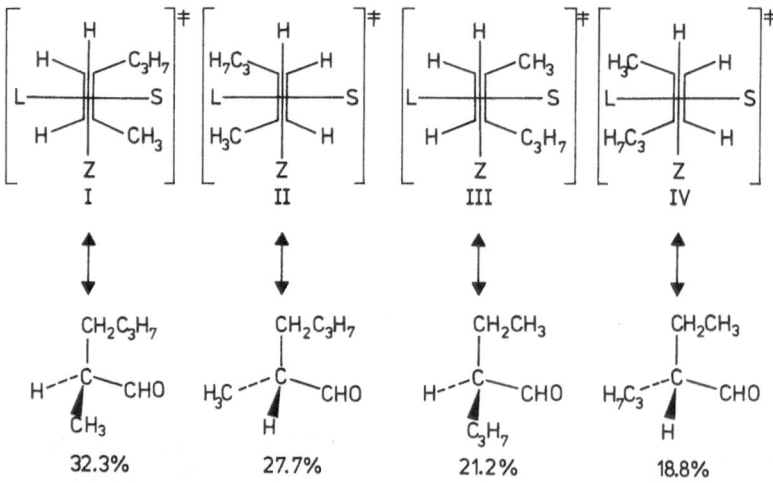

Fig. 9. Transition states determining regio- and stereoselectivity in asymmetric hydroformylation of (Z)-2-hexene using Rh/(—)-DIOP

Table 10. Comparison between predicted prevailing chirality and experimental results in the asymmetric hydroformylation of internal olefins

Catalytic system	Substrate	Chirality of the prevailing enantiomer		Enantiomeric excess (%)	$|G^{\ddagger}_{(III)} - G^{\ddagger}_{(I)}|$[a] cal mol^{-1}
		Predicted	Found		
Rh/(−)-DIOP	(Z)-2-butene	—	(S)	27	327
	(E)-2-butene	(S)	(S)	3.2	47
	(Z)-2-hexene	(S)	(S)[b]	7.6	111
	(E)-2-hexene	(R)	(R)[c]	5.8	85
		(S)	(S)[b]	1.4	20
		(R)	(R)[c]	2.9	42
	(E)-1-phenyl-1-propene	(S)	n.d.	n.d.	—
			(R)[d]	14.4	191
	bicyclo[2,2,1]heptene	(1R, 2R, 4S)	(1S, 2S, 4R)	5.3	62
	bicyclo[2,2,1]octene	(R)	(R)	4.2	61
	2,5-dihydrofurane	(R)	(R)	7.0	83
Rh/((−)-CHIRAPHOS	(Z)-2-butene	—	(S)	18.4	275
	(E)-2-butene	(S)	(S)	18.5	276
	bicyclo[2,2,1]heptene	(1R, 2R, 4S)	(1R, 2R, 4S)	16.4	244
Pt/(−)-DIOP	(Z)-2-butene	—	(S)	14.5	193
	(E)-2-butene	(S)	(S)	24.2	326
	bicyclo[2,2,1]heptene	(1R, 2R, 4S)	(1R, 2R, 4S)	28.4	408
Pt/((−)-CHIRAPHOS	(Z)-2-butene	—	(R)	15.2	226
	(E)-2-butene	(R)	(R)	6.7	99
	bicyclo[2,2,1]heptene	(1S, 2S, 4R)	(1S, 2S, 4R)	8.3	123
Co/(−)-DIOP	(Z)-2-butene	—	(R)	2.4	45
	bicyclo[2,2,2]octene	(S)	[(R)][e]	1.2	20
Ru/(−)-DIOP	(Z)-2-butene	—	(R)	0.5	8
	bicyclo[2,2,2]octene	(S)	[(R)][e]	1.2	20
Pt/((−)-DIOP-DBP	(Z)-2-butene	—	(R)	0.6	9
	(E)-2-butene	(R)	(R)	1.8	27

a Or $|G^{\ddagger}_{(IV)} - G^{\ddagger}_{(III)}|$ respectively (see Figs. 8 and 9); b Referred to 2-methylhexanal; c Referred to-2-ethylpentanal; d Referred to 2-phenylbutanal; e Considerable isomerization of the substrate has been observed [16)

As shown in Table 10,[4] predictions of the chiralities of the prevailing enantiomers fit well, not only for (Z)- but also for (E)-internal olefins. In fact, due to the greater space availability in quadrants Q_1 and Q_2 with respect to quadrants Q_{-1} and Q_{-2} (as shown by the isomeric ratios found in the hydroformylation of (Z)- and (E)-2-hexene), the difference in space availability between quadrants Q_{-1} and Q_{-2} are expected to influence mainly the energy of the activated complexes yielding one or the other antipode.

In the case of aliphatic or alicyclic olefins only norbornene hydroformylation with Rh/(—)-DIOP does not fit into the picture. The contrasting results between (Z)-2-butene and bicyclo[2.2.2]oct-2-ene obtained with Co/(—)-DIOP and Ru/(—)-DIOP catalytic systems have been attributed to extensive isomerization of (Z)-2-butene to (E)-2-butene during hydroformylation [16]. In the case of the only phenyl-substituted substrate investigated, the prediction of the unsaturated carbon atom preferentially formylated is not correct. This type of exceptions is found also for styrene, as it will be discussed later.

4.1.2 Monosubstituted Ethylenes

For monosubstituted ethylenes chiral products arise only when the hydrogen of the catalyst binds to the unsubstituted unsaturated carbon atom of the substrate. Since this product is, in general, the minor one, from the enantiomeric excess no prediction can be made of the face of the substrate preferentially attacked but only the face of the substituted unsaturated carbon atom preferentially formylated to form the chiral product can be established. Indications about all 4 transition states can be obtained using either labelled 2-[^{2}H]-olefins or carrying out a deuterio-formylation instead of a hydroformylation (see Sect. 2.1.5.).

Predictions concerning prevailing chirality, obtained from substrates not containing phenyl groups, are correct and predictions for styrene fail in two cases (Table 11). Predictions concerning the predominating isomer fail in the case of aliphatic substrates only when a heteroatom is directly bound to the double bond. The possible significance of the above failures is discussed in Section 4.2. Over 34 cases wrong predictions have been made in 7 cases, 5 of which concern aromatic substrates. The large values of A (Table 11) frequently found confirm a greater space availability in Q_1 and Q_2 with respect to Q_{-1} and Q_{-2}.

4.1.3 1,1-Disubstituted Ethylenes

The application of the model to 1,1-disubstituted ethylenes is illustrated in Fig. 10, in which R' is bulkier than R. We have assumed, as usual, the following order

---

4 Table 10 as well as the Tables 11 and 12 show the differences in energy of the transition states responsible for enantiomeric and regioisomeric excess. They have been calculated on the basis of the enantiomeric and isomeric ratios. They correspond to the difference in free activation energies ($\Delta\Delta G^+$) for a single-step formation of the metal-alkyl complex intermediate from the substrate and the catalyst complex.

If the metal-alkyl complex intermediates arise from preformed π-complexes which are at equilibrium, the above differences correspond to differences between the sums of the free energy of formation of the π-complexes plus the free energy of activation for the processes leading from the π-complexes to the corresponding alkyl complexes for the enantiomers (or regiosiomers) [H. B. Kagan: Organische Stereochemie, George Thieme Verlag, Stuttgart, 1977, p. 142] (see Sect. 5.2).

Table 11. Comparison between experimental results in asymmetric hydroformylation of monosubstituted ethylenes (with diphosphines as the chiral ligand) and predicted isomeric and enantiomeric composition of the reaction products

Catalytic system	Substrate	Unsaturated carbon atom preferentially attacked by CO		Isomeric excess[b] (%)	A[c,d] cal mol^{-1}	Chirality of the predominating enantiomer		Enantiomeric excess (%)	$\|G^{\ddagger}_{(IV)} - G^{\ddagger}_{(III)}\|$[d] cal mol^{-1}
		Predicted[a]	Found			Predicted[a]	Found		
Rh/(−)-DIOP	1-butene	1	1	85	1490	(R)	(R)	18.8	225
	3-methyl-1-butene	1	1	86	1526	(R)	(R)	15.2	181
	1-pentene	1	1	86	1526	(R)	(R)	19.7	236
	1-octene	1	1	80	1227	(R)	(R)	16.5	197
	3-phenyl-1-propene	1	1	76	1314	(R)	(R)	15.5	206
	styrene	1	2	36	467	(R)	(R)	25.2	319
	3-cyano-1-propene	1	1	10	124	(R)	(R)	21.5	270
	vinyl acetate	1	2	84	1659	(S)	(R)	23.0	318
	N-vinylsuccinimide	1	2	66	1036	(S)	n.r.	19.8	262
Rh/(−)-CHIRAPHOS	1-butene	1	1	8	118	(R)	(R)	7.1	90
	styrene	1	2	42	661	(R)	(R)	24.2	365
Pt/(−)-DIOP	1-butene	1	1	92	2095	(R)	(R)	46.7	668
	1-pentene	1	1	93.4	2227	(R)	(R)	29.1	395
	1-octene	1	1	82	1708	(R)	(R)	18.6	248
	styrene	1	1	28	379	(R)	(S)	28.6	388
Rh/(1R, 3S, 4S)-NMDPP	1-butene	1	1	82	1709	(S)	(S)	44.7	710
	styrene	1	2	24	361	(S)	(R)	45.0	716

[a] On the basis of the model (Fig. 7) and of the results of the hydroformylation of (Z)-2-butene with the same catalytic systems;

[b] Isomeric excess (i.e.) = $\dfrac{\text{prevailing isomer} - \text{minor isomer}}{\text{prevailing isomer} + \text{minor isomer}} \times 100$;

[c] A = $|(G^{\ddagger}_{(II)} + G^{\ddagger}_{(IV)}) - (G^{\ddagger}_{(I)} + G^{\ddagger}_{(III)})|$ (Fig 7);

[d] See Fig. 7

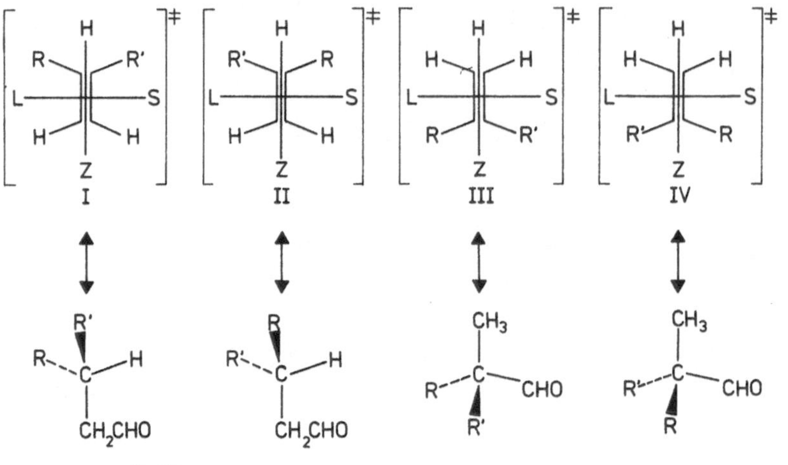

Fig. 10. Transition states determining regio- and stereoselectivity in asymmetric hydroformylation of 1,1-disubstituted ethylenes

of decreasing bulkiness $i\text{-}C_3H_7 > CH_3, n\text{-}C_4H_9 > C_2H_5, C_2H_5 > CH_3, C_6H_5 > CH_3$ $C_6H_5 > C_2H_5$. The results are reported in Table 12. The regioselectivities observed - are in agreement in all cases with those predicted by the model (Fig. 10). Concerning enantioface discrimination, due to the very large prevalence of one chiral aldehyde, the face of the unsaturated carbon atom attacked by hydrogen to form the chiral product and the enantioface preferentially attacked are very likely to coincide.

Due to the relatively small differences in bulkiness between the two substituents at the double bond, and to the relatively large space available in Q_1 and Q_2, a slight difference in the free energies of the activated complexes I and II (Fig. 10) and therefore a small enantiomeric excess is to be expected. Indeed, in most cases, enantiomer excesses are smaller than 25% (corresponding to differences in the free energies of the transition states of less than 400 cal mol^{-1}) and some predictions fail not only for phenyl-substituted but also for aliphatic substrates, especially when the enantiomer excess is very small. Altogether, from 30 data, 5 wrong predictions have been made according to the model.

4.2 Reliability of the Predictions of Isomeric Composition and Enantiomeric Excess in Enantioface-Discriminating Hydroformylation

As shown in the previous section, the reliability of the predictions by the model described in Fig. 6 is surprisingly good in view of the simplicity of the model and of the relatively small isomeric and enantiomeric excesses found in many cases, the differences in energy between the controlling transition states being between 3 and 770 cal mol^{-1}.

Summarizing the data of Table 10, 11 and 12, predictions are wrong in less than 20% of the cases; of the exceptions about 60% concern substrates containing an

Table 12. Comparison between experimental results and predictions in the asymmetric hydroformylation of 1,1-disubstituted ethylenes by different catalytic systems

Catalytic system	Substrate	Unsaturated carbon atom preferentially attacked by CO		Isomeric excess[b]	A[c]	Chirality of the predominating enantiomer		Enantiomeric excess	$|G^{\ddagger}_{(I)} - G^{\ddagger}_{(III)}|$[d]
		Predicted[a]	Found	(%)	cal mol^{-1}	Predicated[a]	Found	(%)	cal mol^{-1}
Rh/(−)-DIOP	2-methyl-1-butene	1	n.r.	n.r.	n.d.	(S)	(R)	0.2	3
	2,3-dimethyl-1-butene	1	1	>98	>3000	(R)	(S)	3.5	52
	2-ethyl-1-hexene	1	1	>96	>2800	(S)	(R)	1.1	16
	2-phenyl-1-propene	1	1	>98	>3000	(R)	(R)	1.6	21
	2-phenyl-1-butene	1	1	>98	>3000	(R)	(R)	1.8	24
Rh/(−)-CHIRAPHOS	2,3-dimethyl-1-butene	1	1	>98	>3000	(R)	(R)	21.4	321
	2-phenyl-1-propene	1	1	>98	>3000	(R)	(R)	21.8	327
Pt/(−)-DIOP	2-methyl-1-butene	1	1	>98	>3000	(S)	(S)	4.2	62
	2,3-dimethyl-1-butene	1	1	>98	>3000	(R)	(R)	19.9	282
	2-phenyl-1-propene	1	1	>98	>3000	(R)	(S)	13.2	196
	2-phenyl-1-butene	1	1	>98	>3000	(R)	(S)	20.7	310
Pt/(−)-CHIRAPHOS	2,3-dimethyl-1-propene	1	1	>98	>3000	(S)	(S)	19.8	296
	2-phenyl-1-propene	1	1	>98	>3000	(S)	(S)	4.0	59
Co/(+)-R*-Sal	2-phenyl-1-propene	1[e]	1	>98	>3000	(S)[e]	(S)	2.8	44
	2-phenyl-1-butene	1[e]	1	~50	—	(S)[e]	(S)	1.4	22

[a] On the basis of the model (Fig. 10) and of the results of the hydroformylation of (Z)-2-butene with the same catalytic systems;   [b] See corresponding note in Table 11;   [c] $A = |(G^{\ddagger}_{(I)} + G^{\ddagger}_{(II)}) - (G^{\ddagger}_{(III)} + G^{\ddagger}_{(IV)})|$ (Fig. 10);   [d] See Fig. 10;   [e] Predicted on the basis of the results of the hydroformylation of 1-butene (according to the model of Fig. 7) with the same catalytic system

aromatic ring conjugated with the olefinic double bond, and 14% concern substrates in which a heteroatom is directly bound to the double bond.

As interactions between phenyl groups of π-complexes of olefins and a phenyl group of the other ligands bound to the metal are known in some square planar platinum (II) complexes [59], it is likely that such a type of attractive interactions hinders the application of our model which is based on repulsive interactions only.

It is interesting that when DIOP-DBP is used as the chiral ligand, due to the rigidity of the dibenzophosphole group, attractive interactions between this ligand and the phenyl groups seem less probable, and the model fits both for vinylic and aliphatic internal olefins, as well as for styrene, no data being available for 2-phenyl-1-propene.

Weak attractive interactions between substrate and metal of the catalytic complex could be responsible for the two exceptions in the prediction of isomeric composition concerning substrates in which a nitrogen atom (N-vinylsuccinimide) or an oxygen atom (vinyl acetate) is directly bound to the olefinic double bond of the substrate. Indeed, in platinum(II) complexes containing vinyl ethers as olefinic ligands, attractive interactions between the platinum atom and the oxygen atom of the olefinic ligand have been found [60].

Particularly interesting are the only exceptions involving non aromatic substrates which have been found for 1,1-disubstituted ethylenes and norbornene with the Rh-DIOP catalytic system (Tables 11 and 12). In the case of open-chain substrates it appears that the larger substituent prefers quadrant Q_1 over quadrant Q_2 (Fig. 6), in contrast to the prediction by the model, whereas in 2-phenyl-1-propene the bulkier phenyl group prefers quadrant Q_2. Since most of the exceptions observed concern substrates containing phenyl groups conjugated with the olefinic double bond, we think that in the Rh/(—)-DIOP system, e.g. because of the effect of groups not directly bound to the metal, space availability is indeed slightly greater in quadrant Q_1 than that in quadrant Q_2, the difference being rather small as shown by the low values of the enantiomeric excess. In this case, 2-phenyl-1-propene and not the aliphatic 1,1-disubstituted ethylenes should be considered as the real exception, in keeping with the behavior of other phenyl substituted olefins.

This hypothesis has been confirmed by the results of deuterioformylation or hydroformylation of deuterated substrates carried out to establish the face preferentially reacting in monosubstituted ethylenes (Table 13).

The results concerning the enantiomeric excess for 1-pentene show that quadrant Q_{-2} is preferred to quadrant Q_{-1}, as predicted both for Pt/(—)-DIOP and Rh/(—)-DIOP catalytic systems. Furthermore, with Pt/(—)-DIOP, quadrant Q_2 is preferred to quadrant Q_1, as predicted, whereas with Rh/(—)-DIOP, quadrant Q_1 is preferred to quadrant Q_2, in agreement with the results obtained with aliphatic 1,1-disubstituted ethylenes. Comparing the results obtained with α-[^{2}H]-styrene and with 2-phenyl-1-propene it appears that the phenyl group prefers quadrant Q_1 (as the n-propyl group in 1-pentene) when ^{2}H is in quadrant Q_2; however, if a methyl group is present, steric repulsion is minimized when methyl occupies quadrant Q_1 and the phenyl group is in quadrant Q_2.

The above considerations concerning the exceptions to our rules show clearly the limits of our model due to the fact that repulsive steric interactions only are considered between substrate and substituents directly bound to the metal atom

Table 13. Asymmetric deuterioformylation (DF) of 1-pentene and styrene and hydroformylation (HF) of α-[^{2}H]-styrene

Substrate	Reaction type	Catalytic system	Reaction product (%)	Prevailing chirality		Enantiomeric excess (%)	Quadrant preferred by the substituent[b]	
							Predicted	Found
1-pentene	DF	Rh/(—)-DIOP	1,3-[^{2}H]$_2$-hexanal	93	(S)	13.1	2	1
			1,3-[^{2}H]$_2$-2-propylpropanal	7	(R)	17.9	−2	−2
1-pentene	DF	Pt/(—)-DIOP	1,3-[^{2}H]$_2$-hexanal	97	(R)	5.4	2	2
			1,3-[^{2}H]$_2$-2-propylpropanal	3	(R)	29.1	−2	−2
α-[^{2}H]-styrene	HF	Rh/(—)-DIOP	3-[^{2}H]-3-phenylpropanal	40	(S)	15.0	2	1
			2-[^{2}H]-2-phenylpropanal	60	(R)	15.0	−2	−2
styrene	DF	Pt/(—)-DIOP	1,3-[^{2}H]$_2$-3-phenylpropanal	64	(R)	10.3[a]	2	1
			1,3-[^{2}H]$_2$-2-phenylpropanal	36	(S)	34.9	−2	−1

[a] Because of isotopic impurities, which cannot be exactly determined, the optical purity is probably between 10.3 and 16.1%
[b] See Figs. 6 and 7

of the catalyst. Furthermore, the substituents bound to the metal are considered as spheres the center of which is situated on the metal-ligand bond. Small differences in space availability of two quadrants (Q_1 and Q_2, Fig. 6) as the ones hypothesized to explain the exceptions observed in the hydroformylation of 1,1-disubstituted ethylenes with the Rh-DIOP catalytic system, affect the reliability of the predictions but not the principles on which the model is based.

In fact, ligands which are not free to rotate around the metal-ligand bond (e.g. diphosphines forming a ring including the metal) cannot be conveniently simulated by spheres. Furthermore, groups belonging to the three ligands considered in our model but not directly bound to the metal could partially occupy the quadrants considered.

4.3 Application of the Model to Enantiomer-Discriminating Hydroformylation

Using the same model and considering the results of the investigation of square planar platinum(II)-chiral olefin complexes [61-63] it is possible to correlate enantioface discriminating hydroformylation with enantiomer discriminating hydroformylation.

The investigation of platinum(II)-chiral olefin complexes has shown that, when the diastereomeric equilibrium is reached, which diastereoface of the olefin is preferentially bound to the metal depends on the type of chirality of the olefin used [61-63]. When an optically active asymmetric ligand is present in the complex and a racemic olefin, is used, one diastereoface will be preferred for complexation and correspondingly one of the antipodes is preferentially complexed [61-63]. Let us suppose that with a certain catalytic system (e.g., Rh/(−)-DIOP), the re-re enantioface of a prochiral α-olefin reacts preferentially. With the same catalytic system the same face of all α-olefins, including the racemic α-olefins, is expected to react preferentially. However, when a racemic olefin is used, two diastereomeric transition states (e.g. a and b in Fig. 11) can form for each of the transition states shown in Fig. 7, depending on which one of the antipodes of the racemic monomer approaches the catalyst.

Interactions between side chain and olefinic carbon atom of the substrate approaching the metal seem to be similar both in the aforementioned platinum(II) complexes and in the above transition states. In fact, in the deuterioformylation of racemic 3-methyl-1-pentene in the presence of an achiral rhodium catalyst it has been shown that the si-si face is preferentially attacked in the (S) antipode whereas,

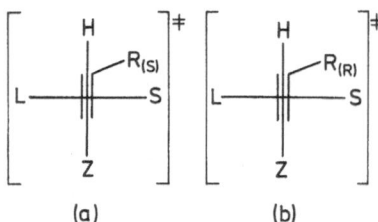

(a)   (b)

Fig. 11a and b. Possible transition state of racemic olefins of course, the re-re face is preferentially attacked in the (R) antipode [64]. Therefore, we can predict that, in the hydroformylation with an asymmetric catalyst which predominantly attacks the re-re face, the transition states involving antipode (R) have a lower free energy than those involving antipode (S) (e.g. in Fig. 11, b has a lower free energy than a), and antipode (R) reacts more rapidly than antipode (S). In this way, enantioface-discriminating and enantiomer-discriminating hydroformylation can be correlated.

As shown in Table 14, starting with the enantioface of an achiral olefin hydroformylated with an asymmetric catalytic system, a correct prediction of the antipode of a racemic substrate mainly hydroformylated with the same catalytic system has been made in the few cases examined up to now.

Table 14. Correlation between enantioface-discriminating and enantiomer-discriminating hydroformylation with Rh/(—)-DIOP catalytic systems[a]

Substrate	Enantioface reacting preferentially		Antipode predicted to form the lower-energy diastereomeric transition state[c]	Antipode reacting preferentially[d]
	Found	Predicted[b]		
1-Pentene	re-re			
(S)-3-methyl-1-pentene[e]	si-si			
(R)-3-methyl-1-pentene[e]	re-re			
(R)(S)-3-methyl-1-pentene	—	re-re	(R)	(R)
(R)(S)-3-methyl-1-hexene	—	re-re	(R)	(R)
(R)(S)-3,4-dimethyl-1-pentene	—	re-re	(R)	(R)
(R)(S)-3,4,4-trimethyl-1-pentene	—	re-re	(S)	(S)
(R)(S)-3-phenyl-1-butene	—	re-re	(S)	(S)

[a] Compare Table 5;
[b] Assuming that the same asymmetric catalytic system preferentially attacks faces having similar geometric requirements in chiral and achiral monosubstituted ethylenes;
[c] On the basis of the experiment mentioned as well as by assuming similar steric interactions between the chiral group and the double bond of the substrate in the postulated transition state (Fig. 11) and in platinum(II) chiral olefin-complexes when the diastereomeric equilibrium is reached [61–63];
[d] Determined after about 50% conversion of the substrate;
[e] From the results of the deuterioformylation of (R)(S)-3-methyl-1-pentene with Rh₂O₃/PPh₃ as the catalyst precursor [64]

5 Possible Significance of the Observed Regularities

In the case of rhodium catalytic systems the observed regularities in hydroformylation have been interpreted on the basis of the currently accepted mechanism for hydroformylation [15]. In the following sections the above explanation is critically reviewed and applied to more recent experiments of asymmetric hydroformylation with different catalytic systems.

5.1 Step Mainly Determining Asymmetric Induction

Hydroformylation is a multistep catalytic process; from the data on the cobalt-catalyzed reaction it has generally been proposed that the reaction occurs according to the scheme in Fig. 12 [21].

Similar schemes have been accepted for the rhodium [57] and the platinum-catalyzed reaction [58]. For cobalt, a possible alternative that, following activation of hydrogen molecules, both hydrogen atoms necessary for hydroformylation are incorporated in the first step [65] was excluded [66].

Although some remarkable discrepancies are evident from the details of the catalytic action of each metal system [3], no fact is known up to now which disagrees with the sequence of the reaction steps or with the nature of the intermediates shown in Fig. 12.

A necessary premise behind all attempts to explain the origin of asymmetric induction and hence the observed regularities is the identification of the step in which mainly asymmetric induction takes place.

Since the two resulting antipodes have the same thermodynamic stability, the asymmetric synthesis must be kinetically controlled. Therefore, the factor controlling the enantiomeric excess must be the difference between the energies of the two diastereomeric transition states leading to one or to the other antipode respectively. Since the process leading from the olefins to the aldehydes is a multistep one, if more steps involving diastereomeric intermediates exist, the two diastereomeric transition states which determine the type of enantiomeric excess in the products must be identified.

From the results of the asymmetric hydroformylation of the isomeric straight-chain butenes we have concluded that asymmetric induction in the case of Rh/(—)-DIOP catalytic complexes does not result either in carbon monoxide insertion or in the following steps described in Fig. 12 but either in the π-complex formation or the alkyl-complex formation [15] (Fig. 13).

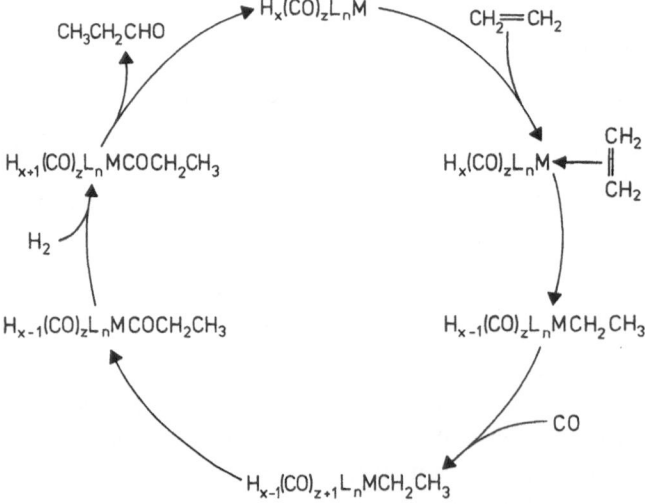

Fig. 12. Proposed hydroformylation mechanism (L = PR_3 or $M_y(CO)_z$ or $M_y(PR_3)_m(CO)_p$)

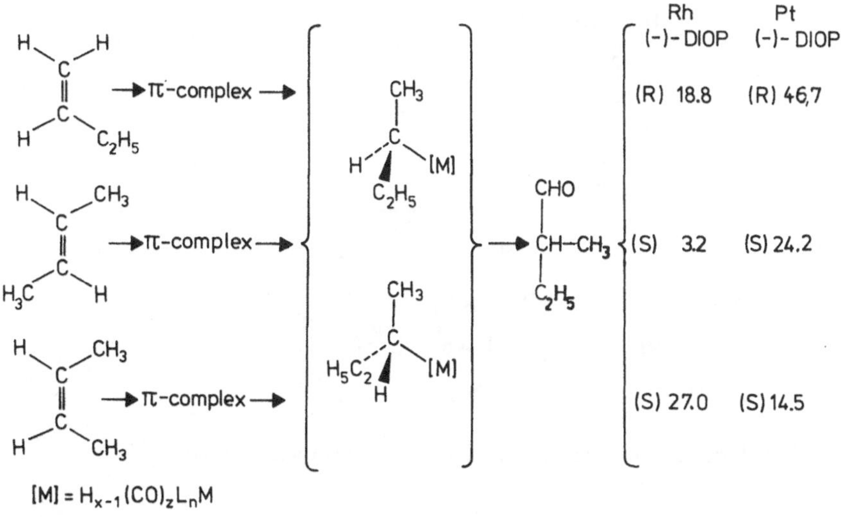

Fig. 13. Asymmetric hydroformylation of linear butenes

The same observation also applies to the Pt/DIOP catalytic system, the contradictory results obtained in preliminary experiments being due to an extensive isomerization of the substrates when high conversion to aldehydes was achieved (see Sect. 2.1.3.) [44].

Therefore, for either antipode, the diastereomeric activated complex controlling optical yield could be either the one corresponding to the formation of the π-complex or the one corresponding to the olefin insertion into the metal-hydrogen bond. In the case of rhodium, it appears from the results of the hydroformylation of 1,2-dimethylcyclohexene and of 2-methylmethylidencyclohexane, that the second case is more probable [10]. In the case of platinum, the fact that isomerization of the substrate, which is very likely to occur via metal alkyl-complex formation, proceeds at a rate similar to or even higher than the hydroformylation rate seems to indicate that the same situation can also be assumed.

The prevalence of the (S) or (R) antipode of the product gives a qualitative idea of the relative energy of the transition states controlling asymmetric induction. However, the values of the enantiomeric excess do not necessarily indicate the extent but only a minimum value of the free energy differences between the above transition states (e.g. the reaction product could racemize after its formation). On this basis the already discussed model for the transition state controlling asymmetric induction has been formulated (Sect. 4.).

As we mentioned earlier (Sect. 2.1.5.), a further complication arises from the fact that, with the exception of the C_{2v} or C_{2h} olefinic substrates, two isomeric reaction products could be formed by *cis* attack of the metal hydride to one face of the prochiral substrate. In principle, if π-olefin complexes are intermediates, the isomeric ratio could be determined in the π-complex formation, two non-interconvertible conformers of each of the two diastereomeric π-complexes being formed. Each conformer then gives rise to a different structural isomer of the reaction products (Fig. 14, paths a, c and a', c').

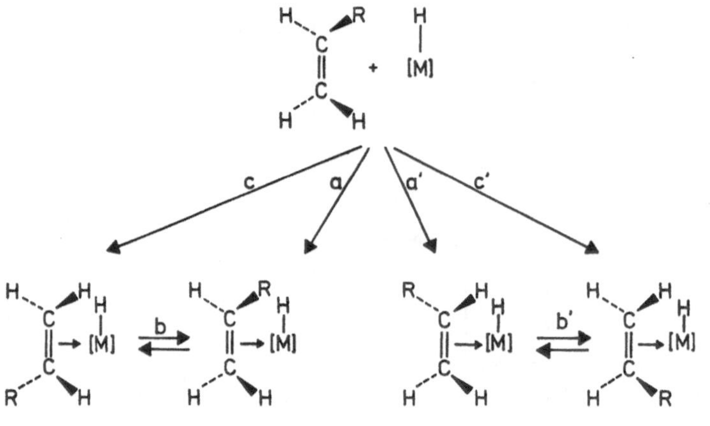

Fig. 14. Possible pathways for the origin of regioselectivity in hydroformylation

However, investigation of square planar (olefin) Pt(II)complexes has shown that the energy barrier to the interconversion of the rotamers can be low [67]; therefore, it is possible (Fig. 14, paths a + b and a' + b') that for each diastereomeric π-complex only one rotamer is formed in the interaction between catalyst and olefin and that the rotameric equilibrium is reached very rapidly.

In any case, if the step corresponding to the metal alkyl-complex formation is indeed the first irreversible step in asymmetric hydroformylation, both enantiomeric excess and isomeric ratio are very likely to be regulated, in the case examined, in this reaction step and should be considered together in any attempt to explain asymmetric induction.

5.2 Factors Influencing Activation Energy Differences in the Step Determining Asymmetric Induction

The simplest case which can be used to explore the factors influencing the difference in the energies of the diastereomeric transition states, which determine asymmetric induction, is the hydroformylation of (Z)-2-butene with the Rh/(−)-DIOP or Pt/(−)-DIOP catalytic system. In this case, the asymmetric induction cannot be connected with enantioface discrimination in the step leading to the π-complex because this olefin has no enantiofaces [68]. In the first step of the reaction it is assumed that a π-complex is formed by interaction between substrate and catalyst. This π-complex, depending on its geometry, can exist in two different conformations arising from the rotation of the olefin around the metal-olefin π-system-bond axis.

Investigation of π-olefin metal complexes with (Z)-2-butene or similar C_{2v} olefins has shown that the energy barrier to the rotation is rather low [67,69,70] and that therefore the two conformers must rapidly reach the equilibrium even at room temperature.

It is therefore likely that for the catalyst-(Z)-2-butene complex two rapidly interconverting conformers exist. The olefin moieties in both conformers can then undergo insertion into a M—H bond, thus resulting in the formation of two diastereomeric sec-butyl complexes, each of which would lead to the corresponding antipode of 2-methylbutanal.

As discussed in the previous section the activated complexes transformed via the π-olefin complexes to non-interconvertible sec-butyl complexes probably have the highest energy among all the transition states involved in the multistep catalytic reaction.

The difference between the energies of the above diastereomeric activated complexes leading to the two antipodes in the simplest case determines at least qualitatively the type of asymmetric induction. This difference can be related [15] both to the relative stabilities of the two conformers of the π-complex and to the activation energies leading from each of the conformers to the corresponding activated complexes.

Two interesting cases may be considered:

a) $(G_2^{\ddagger} - G_1^{\ddagger}) = (G_2^0 - G_1^0)$. In this case, the activation energy for the step π-complex→metal alkyl is the same for the two conformers and the predominant antipode arises from the more stable π-complex. The enantiomeric excess therefore reflects the relative stabilities of the two conformers (Fig. 15a)[5].

b) $G_1^0 < G_2^0$; $G_2^{\ddagger} < G_1^{\ddagger}$. In this case, the activation energy for the step π-complex→metal alkyl is lower for the conformer having the higher energy of formation and it is $|\Delta G_2^{\ddagger} - \Delta G_1^{\ddagger}| > |\Delta G^0|$. The prevailing antipode arises from the less stable π-complex (Fig. 15b).

A possible example for case a is given in Fig. 16. π-Complexes and transition states are very similar and the activation process consists in a translation of the olefinic bond that leads one of the unsaturated carbon atoms nearer to hydrogen. Activation energies of the transition states are supposed to be very similar starting with either conformer Ia or IIa.

A possible representation of case b is given in Fig. 17. In the π-complexes, the double bond axis of the substrate is not coplanar with the M—H bond and the

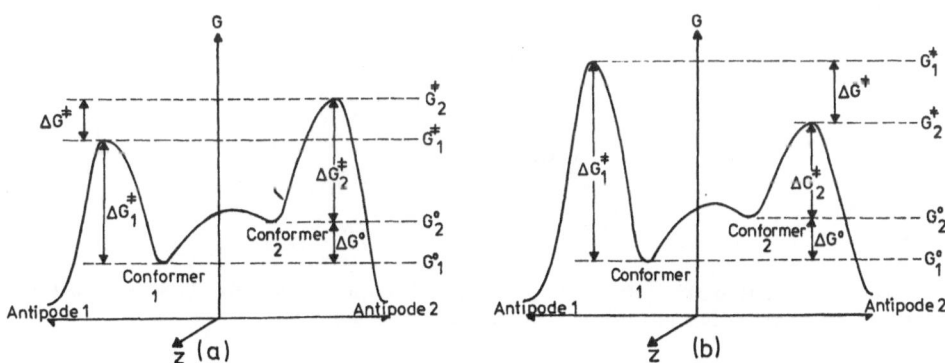

Fig. 15a and b. Possible kinetic models for asymmetric hydroformylation of (Z)-2-butene

---

5 When the lowest activation energy corresponds to the more stable conformer of the π-complex, the enantiomeric excess will reflect only qualitatively the relative stability of the two conformers

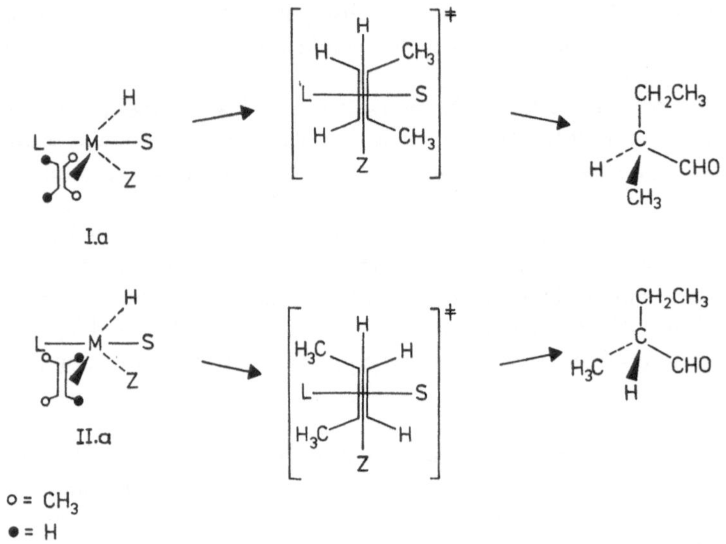

o = CH$_3$
● = H

Fig. 16. Possible relationship between π-complexes and transition states for the kinetic model *a* of Fig. 15 (For the reaction products composition see Tables 2 and 4)

o = CH$_3$
● = H

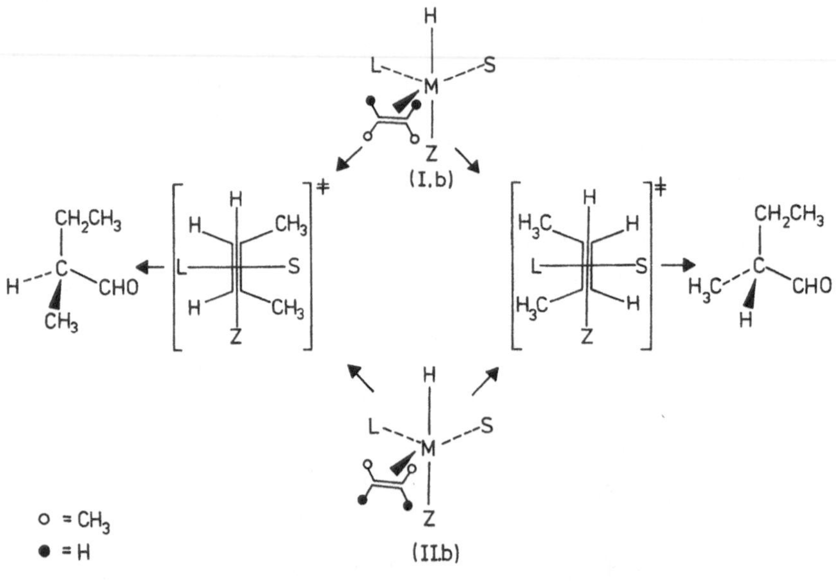

Fig. 17. Possible relationship between π-complexes and transition states for the kinetic model *b* of Fig. 15 (For the reaction products composition see Tables 2 and 4)

activation process, before translation, as described in case *a*, involves rotation of the double bond to reach the geometry which is necessary for insertion. The activation energy is lower starting with the less stable isomer I b than with the more stable isomer II b to form the prevailing enantiomer.

116

In the same way, enantiomer-discriminating hydroformylation can be discussed.

Unfortunately, the isomeric composition and the enantiomeric excess found in the hydroformylation products give only information on the relative free energy of the transition states and does not allow a distinction to be made between the two kinetic models *a* and *b*. However, the regularities observed (Sect. 4.1.) indicate that the same kinetic model (*a* or *b*) must be valid in most (if not in all) of the cases investigated.

5.3 Possible Mechanism for Asymmetric Induction in the Step Corresponding to the Formation of the Metal Alkyl Complex

As shown in the asymmetric hydroformylation of 1- and 2-butene with Rh/(—)-DIOP or Pt/(—)-DIOP catalytic systems, it seems likely that asymmetric induction occurs mainly in the step in which the π-olefin complexes, which are assumed to be formed in the first reaction step, are transformed into the corresponding metal-alkyl intermediate [15].

The results obtained with 1- and 2-butene can be extended to other olefinic hydrocarbons used as the substrate.

The good results obtained in the prediction of isomeric composition and of enantiomeric excess using the model described in Section 4, indicate that the difference in free energy of the transition states in the cases examined is mainly related to repulsive interactions between substrate and catalyst.

Since π-olefin-complexes are relatively well-known compounds, it would be interesting in order to improve our understanding of the asymmetric induction mechanism, to relate the relative stabilities of the diastereomeric π-complexes and of their conformers to the structure of the transition states. An attempt along this line has been made [15] using the assumption that the activation energy to reach the transition states starting with the corresponding diastereomeric π-olefin complexes is very similar. In this case, at least the sign of the energy difference between the diastereomeric transition states corresponds to the sign of the difference between the free energy of formation of the corresponding π-complexes. Unfortunately, this assumption could never be verified of falsified. Some indications in this sense are given by the results of the hydroformylation of norbornene, in which almost exclusive formation of the exo-isomer indicates that the less hindered diastereoface, which should be preferentially complexed [71], is hydroformylated. However, in asymmetric hydrogenation some NMR evidence exists [72] that the more stable diastereomeric π-olefin complex reacts less rapidly than the less stable one. In view of some similarities in the first two steps of hydrogenation [73, 74] and hydroformylation [21], this result might suggest that a detailed structure for the π-olefin complexes cannot be proposed on the basis of our model for the transition states.

The model we have used for the description of the transition states implies that also the corresponding π-olefin complexes contain an asymmetric metal atom and that changes of configuration at the metal occur more slowly than the intramolecular transformation of the π-complex into the metal-alkyl complex.

The same four substituents around the metal in the transition state should exist also in the corresponding π-complexes. The main open problems concerning the

117

π-complexes are, however, the relative position of the hydrogen atom and of the double bond (Fig. 16, Ia and Fig. 17 Ib) and the actual geometry of the complex. π-Olefin complexes containing a hydride ligand are known having a trigonal bipyramidal [75] or a tetrahedral structure [76].

The nature of the substituents L, S, and Z mentioned in the model for the transition state, if a monometallic catalyst is postulated, must be a PR_3 group or a CO group; if plurimetallic complexes are postulated, $M_y(CO)_x$ or $M_y(PR_3)_n (CO)_{x-n}$ could also be a ligand.

The geometry of the complex seems to be of paramount importance in the determination of repulsive interactions in the different quadrants around the metal. The most significant examples are given by the deuteroformylation of monosubstituted ethylenes in which the aliphatic substituent occupies in the model of the transition state (Fig. 6) preferentially quadrant Q_2 in the Pt/(−)-DIOP catalytic system but quadrant Q_1 in the Rh/(−)-DIOP catalyst although, according to our model, the chirality at the metal atom, as determined by the hydroformylation of (Z)-2-butene, is the same.

The geometry of the complexes might also be important in favouring attractive interactions between phenyl groups of the ligands and of the substrate which seem to be responsible for the anomalous behavior, on the basis of the model chosen, of the substrates in which the double bond is conjugated with a phenyl group. However, other alternative explanations are possible for this phenomenon and are currently under investigation.

6 Conclusion

Among the great number of hydrocarbonylation reactions occurring according to Scheme 3, only the hydroalkoxycarbonylation reaction (X = OR) has been investigated up to now.

The best optical yields obtained in the hydroformylation are comparable with the highest yield obtained in hydroalkoxycarbonylation using Pd catalysts. In Table 15 the results obtained in hydroformylation with rhodium or platinum catalysts are compared with those obtained in hydroalkoxycarbonylation using identical substrates and identical optically active ligand [9].

The same regularities observed in hydroformylation seem to exist also in hydroalkoxycarbonylation, the prevailing chiralities in the rhodium-catalyzed hydroformylation being opposite to those in the palladium-catalyzed hydroalkoxycarbonylation. These regularities show that the model used to interpret the experimental hydroformylation data might also be applied to other hydrocarbonylation reactions.

$$\text{C=C} + HX + CO \longrightarrow H{-}C{-}C{-}COX$$

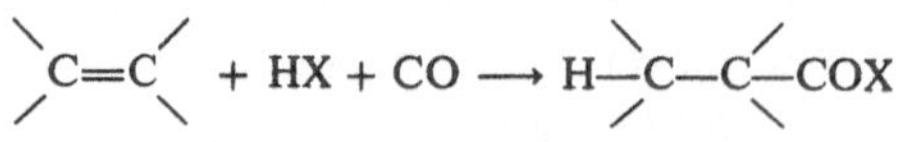

X = H, OH, OR, NH_2, NHR, SH

Scheme 3

Table 15. Influence of the structure of the substrate on the prevailing chirality and on the maximum optical yield obtained in asymmetric hydrocarbonylation with different metallic components of the catalyst in the presence of the same asymmetric ligand [(−)-DIOP]

Substrate	Hydroformylation RhH(CO)(PPh₃)₃ [a]		Hydroformylation [(−)-DIOP] PtCl₂/SnCl₂ [a]		Hydrocarboxylation PdCl₂ [a]	
	Prevailing chiralty	Optical yield (%)	Prevailing chiralty	Optical yield (%)	Prevailing chiralty	Optical yield (%)
1-butene	R	18.8	R	52.6	S	20
1-pentene	R	19.4	R	3.5		
1-octene	R	16.5	R	4.2		
3-methyl-1-butene	R	15.2			S	10.3
3,3-dimethyl-1-butene					S	2
styrene	R	25.2	S	12.2	S	10
allylbenzene	R	15.5				
cis-butene	S	27.0	S	9.3	R	20.7
trans-butene	S	3.2	S	14.4	S	23.2
norbornene	(1S, 2S, 4R)	5.3	(1R, 2R, 4S)	28.4	(1R, 2R, 4S)	4.5
2-methyl-1-butene	R	1.0	S	4.2	R	4.3
2,3-dimethyl-1-butene	S	3.5	R	19.9	S	4.6
2,3,3-trimethyl-1-butene					S	19.5
α-methylstyrene	R	1.6	S	13.2	S	59
α-ethylstyrene	R	1.8	S	20.7	S	49

[a] Compound used as the catalyst precursor

119

The fact that with (Z)-2-butene opposite prevailing chiralities are obtained in hydroformylation, using Rh/(—)-DIOP or Pt/(—)-DIOP catalytic systems, or in hydroalkoxycarbonylation, using the Pd/(—)-DIOP catalytic system, can be explained according to the model assuming that the metal atoms have opposite chiralities in the two cases.

Considering the experiments of hydroalkoxycarbonylation [9], the number of asymmetric hydrocarbonylation results to which the model discussed in Section 4 has been applied is about 200. The wrong predictions are 15%, 83% of which concern experiments with aromatic substrates. Even if the reasons we have given to justify the success of the model may be wrong, the model is undoubtedly very useful to analyze the results of asymmetric hydrocarbonylation with different catalytic systems.

The relatively low enantiomeric excess achieved up to now in hydroformylation does not prove that the synthetic potential of this reaction in the synthesis of optically active oxygenated products is low. In contrast to the research on asymmetric hydrogenation which has mainly been directed to the synthesis of a single class of compounds of practical importance, no comparable effort has been made to synthesize one single compound with high optical purity through hydrocarbonylation.

By comparing the results of asymmetric hydroformylation with those of asymmetric hydrogenation of unsaturated hydrocarbons [4], it appears that indeed the enantiomeric excesses obtained are not very different.

A comparison of the results obtained in hydroformylation with those of other not exhaustively investigated reactions [4], such as hydrosilylation, cross coupling, carbenoid reactions, shows that the necessity to achieve a steric control in the great number of catalytic steps under relatively severe temperature and pressure conditions considerably retards the progress in the field.

An important result achieved so far through the research on asymmetric hydroformylation is an improvement in our basic understanding of the catalytic asymmetric syntheses carried out in the presence of soluble transition metal complexes.

The great number of regularities observed is indeed surprising in view of the small differences in energy between the activated complexes responsible for asymmetric induction, as indicated by the low enantiomeric excesses generally observed. However, the possibility to formulate a model which, on the basis of steric interactions, correctly predicts for a larger number of cases the type of enantiomer that predominates in the products, shows that in the case of asymmetric hydroformylation, other unpredictable factors (e.g. difference in solvation of the transition states) do not substantially affect the free energy difference between the diastereomeric transition states. Furthermore, one can use with confidence substrates having different structures to investigate the structure and the stereochemistry of the catalytic complexes which have not yet been directly characterized in detail.

Finally, the above research has resulted in a better understanding of the hydroformylation mechanism, particularly of the first substantially irreversible step of the reaction and, consequently, has allowed us to formulate a possible model for the regulation of the isomeric composition of the hydroformylation products which, on the basis of our experiments, seems to be mainly connected with steric factors.

7 Acknowledgements

We thank the 'Schweizerischer Nationalfonds zur Förderung der wissenschaftlichen Forschung' for their financial support.

We express our gratitude to Miss B. Straub for revising the english form of our manuscript.

8 Appendix: Optical Purities of the Ligands Considered in this Review

R*-SAL	$[\alpha]_D^{25} = +183$ for the (S) enantiomer. Optical purity $\sim 100\%$ based on the optical purity of the starting (S)-1-phenylethylamine [6].
DIOP	$[\alpha]_D^{22} = -12.6$ for the (R,R) enantiomer. Optical purity is higher than 92% which is the maximum optical yield obtained in the rhodium-catalyzed asymmetric hydrogenation with this ligand [77].
DIOP-DBP	$[\alpha]_D^{22} = -65.5$ for the (R,R) enantiomer. Optical purity should be similar to that of DIOP because both ligands have the same precursor [78].
CHIRAPHOS	$[\alpha]_D^{27} = -211$ for the (S,S) enantiomer. Optical purity $\sim 100\%$, corresponding to the maximum optical yields obtained in the rhodium-catalyzed asymmetric hydrogenation with this ligand [79].
CBDPP	Optical rotation not reported [39]. $[\alpha]_D^{22} = -18.6$ for the (R,R) enantiomer [80]. An optical yield of 91% has been reported for the rhodium-catalyzed asymmetric hydrogenation with this ligand [81].
CBDBP	Optical rotation not reported [39]. This ligand and CBDPP have the same precursor.
CHDPP	Optical rotation not reported [39]. $[\alpha]_D^{22} = -52.8$ for the (R,R) enantiomer [80]. The highest optical yield reported for the rhodium-catalyzed asymmetric hydrogenation is 39% [80].
CHDBP	Optical rotation not reported [39]. This ligand and CHDPP have the same precursor.
CHDPPO	$[\alpha]_D + 43.3$ for the (S,S) enantiomer. The optical purity of the starting diol was 85% [39]. The highest optical yield reported for the rhodium-catalyzed asymmetric hydrogenation is 78.9% [82].
BzMePhP*	The (+) enantiomer has the absolute configuration (R). Optical rotation not reported. Optical purity is presumably high based on the preparation of the ligand [34].
CAMP	Optical rotation not reported [32]. Optical purity is presumably high, based on the results of the rhodium-catalyzed asymmetric hydrogenation reported from the same research laboratory [83].
M(2MB)PP	Optical rotation not reported [8]. Optical purity unknown.
MePr^nPhP*	The (−) enantiomer has the absolute configuration (R). Optical rotation not reported [8]. Optical purity unknown.

NMDPP   The (1R, 3S, 4S) epimer has positive optical rotation. Optical rotation not reported [34]. The highest optical yield reported in the rhodium-catalyzed asymmetric hydrogenation is 65% [84].

PAMP   Same remarks as for CAMP.

9 References

1. Falbe, J.: J. Organometal. Chem. *94*, 213 (1975)
2. Roelen, O.: Ger. Pat. 849, 548 (1938); Chem. Zentr. *1953*, 927
3. Pino, P.: J. Organometal. Chem., *200*, 223 (1980)
4. Pino, P., Consiglio, G., in: Fundamental Research in Homogeneous catalysis. Tsutsui, M., Ugo, R. (eds.). New York, London: Plenum Publ. Corp. 1977
5. Himmele, W. et al.: D.O.S. 2.132.414 (1971); C.A. *78*, 97328 (1973)
6. Botteghi, C., Consiglio, G., Pino, P.: Chimia *26*, 141 (1972)
7. Pino, P. et al.: J. Chem. Soc. C *1971*, 1640
8. Siegel, H., Himmele, W.: Angew. Chem. *92*, 182 (1980)
9. Consiglio, G., Pino, P.: Adv. Chem. Ser., submitted for publication
10. Stefani, A. et al.: J. Amer. Chem. Soc. *99*, 1058 (1977)
11. Izumi, Y., Tai, A.: Stereodifferentiating Reactions. Tokyo: Kodansha Ltd.; New York: Academic Press 1977
12. Pino, P., Stefani, A., Consiglio, G., in: Catalysis in Chemistry and Biochemistry. Theory and Experiment. Pullman, B. (ed.), p. 347. Dordrecht: D. Reidel Publishing Co. 1979
13. Morrison, J. D., Mosher, H. S.: Asymmetric Organic Reactions. Englewood Cliffs, N. J.: Prentice-Hall, Inc. 1971
14. Stefani, A., Tatone, D., Pino, P.: Helv.. Chim. Acta 59, 1649 (1976)
15. Pino, P. et al.: Adv. Chem. Ser. *132*, 295 (1974)
16. Piacenti, F. et al.: Chim. Ind. (Milan) *60*, 808 (1978)
17. Botteghi, C. et al.: J. Organometal. Chem. *161*, 197 (1978)
18. Mignani, G., Patin, H., Dabard, R.: J. Organometal. Chem. *169*, C19 (1979)
19. Botteghi, C., Consiglio, G., Pino, P.: to be published
20. Feder, H. M., Halpern, J.: J. Amer. Chem. Soc. *97*, 7186 (1975)
21. Pino, P., Piacenti, F., Bianchi, M., in: Organic Syntheses via Metal Carbonyls. Wender, I., Pino, P. (eds.). New York: J. Wiley 1977
22. Tanaka, M. et al.: Chem. Lett. *1972*, 483
23. Ogata, I., Ikeda, Y.: Chem. Lett. *1972*, 487
24. Salomon, C. et al.: Chimia *27*, 215 (1973)
25. Tanaka, M., Ikeda, Y., Ogata, I.: Chem. Lett. *1975*, 1115
26. Consiglio, G. et al.: Angew. Chem. *85*, 665 (1973)
27. Botteghi, C., Branca, M., Saba, A.: J. Organometal. Chem. *184*, C17 (1980)
28. Consiglio, G., Morandini, F., Pino, P.: manuscript in preparation
29. Watanabe, Y. et al.: Bull. Chem. Soc., Japan *52*, 2735 (1979)
30. Botteghi, C.: Gazz. Chim. Ital. *105*, 233 (1975)
31. Fritschel, S. J. et al.: J. Org. Chem. *44*, 3152 (1979)
32. Tinker, H. B., Solodar, A. J.: Can. Pat. 1.027.141 (1973); C.A. *89*, 42440m (1978)
33. Botteghi, C. et al.: Chim. Ind. (Milan) *60*, 16 (1978)
34. Tanaka, M. et al.: Bull. Chem. Soc. Japan *47*, 1698 (1974)
35. Salomon, C.: Dissertation, ETH Zürich 1975
36. Botteghi, C., Salomon, C.: Tetrahedron Lett. *1974*, 4285
37. Becker, Y., Eisenstadt, A., Stille, J. K.: J. Org. Chem. *45*, 2145 (1980)
38. Botteghi, C., Consiglio, G., Pino, P.: Ann. Chem. *1974*, 864
39. Hayashi, T. et al.: Bull. Chem. Soc. Japan *52*, 2605 (1979)
40. Bayer, E., Schurig, V.: Chem. Techn. *1976*, 212
41. Pino, P. et al.: D.O.S. 2.359.101 (1973); C.A. *81*, 90644 (1974)

42. Hsu, C.-Y.: Dissertation, Cincinnati 1974; C.A. 82, 154899 (1975)
43. Haelg, P., Consiglio, G., Pino, P.: manuscript in preparation
44. Haelg, P., Consiglio, G., Pino, P.: Proceeding of the 2nd International Symposium on Homogeneous Catalysis, September 1–3, 1980, Düsseldorf, p. 22
45. Consiglio, G., Pino, P.: Helv. Chim. Acta 59, 642 (1976)
46. Consiglio, G., Arber, W., Pino, P.: Chim. Ind. (Milan) 60, 396 (1978)
47. Consiglio, G., Pino, P.: Isr. J. Chem. 15, 221 (1976/77)
48. Kawabata, Y., Suzuki, T. M., Ogata, I.: Chem. Lett. 1978, 361
49. Pregosin, P. S., Sze, S. N.: Helv. Chim. Acta 61, 1848 (1978)
50. G. Erre: Thesis, Sassari 1978
51. Stefani, A., Tatone, D.: Helv. Chim. Acta 60, 518 (1977)
52. Scalone, M.: unpublished results
53. Abley, P., McQuillin, F. J.: J. Chem. Soc. (C) 1971, 844
54. Natta, G.: Ric. Scientifica Suppl. 1958, 28
55. Pino, P., Consiglio, G., in: Fundamental Research in Homogeneous Catalysis. Tsutsui, M. (ed.), p. 519. New York, London: Plenum Press 1978
56. Shurig, V.: J. Mol. Catal. 6, 75 (1979)
57. Evans, D., Osborn, J. A., Wilkinson, G.: J. Chem. Soc. (A) 1968, 3133
58. Orchin, M.: Proceedings of the Symposium on New Hydrocarbon Chemistry. San Francisco: August 29–September 3, 1976, p. 482
59. Ball, R. G., Payne, N. C.: Inorg. Chem. 15, 2494 (1976)
60. Lazzaroni, R., Mann, B. E.: J. Organometal. Chem. 164, 79 (1979)
61. Lazzaroni, R., Salvadori, P., Pino, P.: Chem. Comm. 1970, 1164
62. Lazzaroni, R. et al.: J. Organometal. Chem. 99, 475 (1975)
63. Salvadori, P., Lazzaroni, R., Bertozzi, S.: Proceedings of the IX Convegno Nazionale di Chimica Organica, Salsomaggiore, Italy, 1976, p. 11
64. Stefani, A., Tatone, D., Pino, P.: Helv. Chim. Acta 62, 1098 (1979)
65. Niwa, M., Yamaguchi, M.: Shokubai 3, 264 (1961)
66. Consiglio, G., Oldani, F., Pino, P.: manuscript in preparation
67. Ashley-Smith, J., — et al.: J.C.S. Dalton 1974, 128
68. Paiaro, G. et al.: Makromol. Chem. 71, 184 (1964)
69. Alt, H. et al.: J. Organometal. Chem. 102, 491 (1975)
70. Koemm, U., Kreiter, C. G., Strack, H.: J. Organometal. Chem. 148, 179 (1978)
71. Fischer, K. et al.: Angew. Chem. 85, 1002 (1973)
72. Brown, J. M., Chalomer, P. A.: J.C.S. Chem. Comm. 1980, 344
73. Chan, A. S. C., Pluth, J. J., Halpern, J.: J. Amer. Chem. Soc. 102, 5952 (1980)
74. Chan, A. S. C., Halpern, J.: J. Amer. Chem. Soc. 102, 838 (1980)
75. Klein, H.-F.: Angew. Chem. 92, 362 (1980)
76. Klazinga, A. H., Teuben, J. H.: J. Organometal. Chem. 157, 413 (1978)
77. Kagan, H. B. et al.: Bull. Soc. Chim. Belg. 88, 923 (1979)
78. Dang, T. P., Poulin, J. C., Kagan, H. B.: J. Organometal. Chem. 91, 105 (1975)
79. Fryzuk, M. D., Bosnich, B.: J. Amer. Chem. Soc. 99, 6262 (1977)
80. Aviron-Violet, P., Colleuille, Y., Varagnat, J.: J. Mol. Catal. 5, 41 (1979)
81. Glaser, R., Blumenfeld, J., Twaik, M.: Tetrahedron Lett. 1977, 4639
82. Tanaka, M., Ogata, I.: J.C.S. Chem. Comm. 1975, 735
83. Knowles, W. S., Sabacky, M. J., Vineyard, B. D.: Adv. Chem. Ser. 132, 274 (1974)
84. Morrison, J. D. et al.: J. Amer. Chem. Soc. 93, 1301 (1971)
85. Hanson, R. K.: J. Amer. Chem. Soc. 88, 2731 (1966)
86. Thorn, D. L., Hoffmann, R.: J. Amer. Chem. Soc. 100, 2079 (1978)
87. Dedieu, A.: Inorg. Chem. 20, 2803 (1981)

The Group 5 Heterobenzenes
Arsabenzene, Stibabenzene and Bismabenzene

Arthur James Ashe, III

Department of Chemistry, The University of Michigan, Ann Arbor, Michigan 48 109, USA

Table of Contents

The arsabenzene ring system has been actively studied for just over a decade. Its bond delocalization, diamagnetic ring current and electronic structure demonstrate that arsabenzene has a high degree of aromatic character. While its lack of basicity strongly differentiate it from pyridine, arsabenzene has a rich organic chemistry quite similar to that of normal benzocyclic aromatics.

Stibabenzene and bismabenzene have been much less studied due to their greater lability. This higher reactivity coupled with spectroscopic evidence suggests that aromatic character decreases with increasing atomic number. However, the spectra of the group 5 heterobenzenes strongly emphasize similarities in the series. Thus, the entire family of group 5 heterobenzenes is closely related to their benzocyclic cousins.

1 Introduction

Until recently only a handful of main group elements — carbon, nitrogen, oxygen and sulfur — were known to be able to form stable (p-p)π bonds. The complete lack of such bonds in compounds of the heavier elements had been rationalized by theoretical treatments by Pitzer [1] and Mulliken [2]. Indeed, many commonly used textbooks of inorganic chemistry contain statements such as, "... *p-p-π bonding is of little importance in the third and higher period elements, ...*" [3].

A fundamental breakthrough came in 1966 when G. Märkl succeeded in preparing 2,4,6-triphenylphosphabenzene *1*. [4] The compound was a surprisingly stable yellow crystalline material, which of necessity contained a phosphorus-carbon (3p-2p)π bond. Presumably any inherent instability of the bond was compensated for by the aromatic stabilization of the phosphabenzene ring. Subsequently, both Märkl [5,6] and Dimroth [7] in Germany and Bickelhaupt [8] in The Netherlands prepared a variety of heavily substituted phosphabenzenes. Following our synthesis of the whole family of group 5 heterobenzenes, phosphabenzene *2* [9], arsabenzene *3* [9], stibabenzene *4* [10] and bismabenzene *5* [11], it became clear that even the heaviest elements are able to take part in (p-p) π aromatic bonding.

Together with the parent compound, pyridine, the group 5 heterobenzenes comprise a unique series in which elements of an entire column of the periodic table have been incorporated into aromatic rings. The comparative study of this series seems particularly interesting and should clarify the concepts of aromaticity and of element-carbon π-bonding.

Although several comprehensive reviews have appeared on phosphabenzene [5-8, 12-14], the heavier heterobenzenes [6, 8, 12, 13, 15] have been only modestly covered in the secondary literature. The present work attempts to review arsabenzene, stibabenzene and bismabenzene. Phosphabenzene and pyridine chemistry are only selectively treated for comparison.

2 Synthesis

2.1 9-Heteroanthracenes

The synthesis of the first arsabenzene, 9-arsaanthracene, *6*, was simultaneously communicated by Bickelhaupt [16] and Jutzi [17] in 1969. The elimination of hydrogen chloride from 9,10-dihydroaarsanthracene *7* afforded *6*. These precursors are available from either reduction of the corresponding arsinic acids *8* [16, 18] or by the exchange reaction of dihydrostannaanthracenes *9* with arsenic trihalides [17].

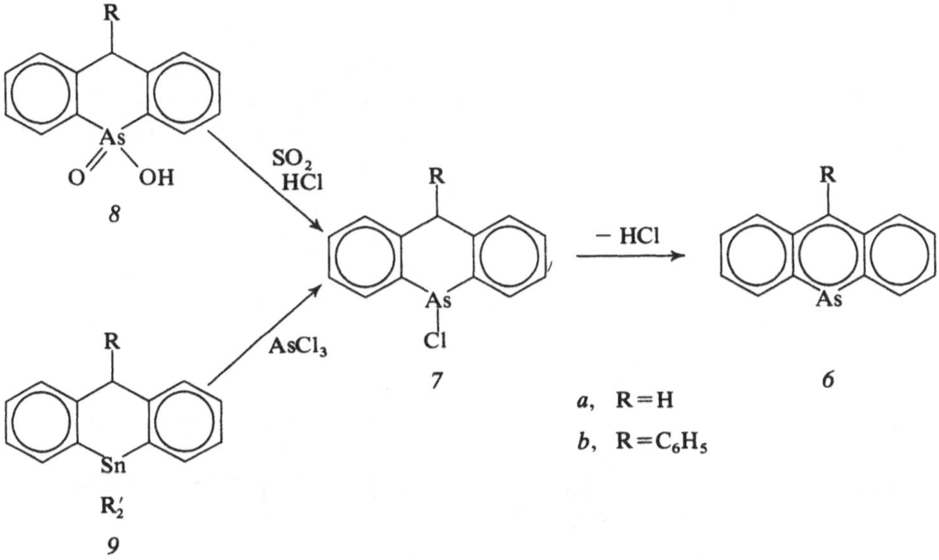

a, R=H

b, R=C$_6$H$_5$

The highly labile 9-arsaanthracene itself may be characterized spectroscopically in dilute solution, but apparently forms dimer *10* and higher polymers on attempted isolation [19]. However, 10-phenyl-9-arsaanthracene *6b* is much more stable and has been isolated as a crystalline solid [20].

Analogous syntheses of 9-phosphaanthracenes have been accomplished [8, 21, 22]. However, attempted extension to 9-stibaanthracene *11* was less satisfactory. On treatment of 9-chloro-9,10-dihydrostibaanthracene *12* with base, the stibaanthracene could not be detected spectroscopically. However, the isolation of the dimer *13* of stibaanthracene is consistent with its intermediacy [19].

7a, E=As
12, E=Sb

6a, E=As
6a, E=Sb

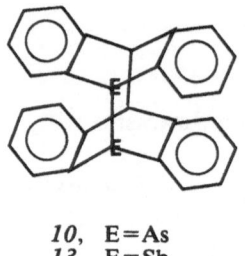

10, E=As
13, E=Sb

2.2 Unsubstituted Heterobenzenes

The successful conversion of *9* to *6* provided the original inspiration for the synthesis of the parent arsabenzene *2*. The stannohydration of 1,4-pentadiyne *14* with dibutyl-tin dihydride furnished the necessary organotin precursor *15* [23]. The exchange reaction of 1,4-dihydro-1,1-dibutylstannabenzene *15* with arsenic trichloride gave arsabenzene in a single step since *16* loses hydrogen chloride on warming or addition of base [9]. An analogous synthesis afforded phosphabenzene *2*. Like most arsines arsabenzene is oxygen-sensitive, but it may be conveniently handled under an inert atmosphere. It is a distillable liquid, which does not react with water, mild acids or bases and is thermally stable to >200 °C.

This synthesis has been extended to the heavier heterobenzenes. The exchange reaction of *15* with antimony trichloride gave 1-chloro-1,4-dihydrostibabenzene *18* as a crystalline solid [10]. On treatment with base, *18* afforded stibabenzene *4*. While stibabenzene can be isolated and characterized, it rapidly polymerizes even from dilute solutions at ambient temperature. The reaction of bismuth trichloride and *15* afforded a good yield of crystalline 1-chloro-1,4-dihydrobismabenzene *19*, which on reaction with base at 0 °C gave only intractable material [11]. However, bismabenzene *5* has been detected both spectroscopically [24] and *via* chemical trapping [11].

2.3 Substituted Heterobenzenes

Although stannohydration of 1,4-diynes *20* of necessity leads to heterobenzenes which are unsubstituted in the 3- and 5-positions, the method can be used to prepare 2- and 4-substituted heterobenzenes. A variety of 2-alkylarsabenzenes *23a–d* have been prepared [25, 26]. Thus 1,4-hexadiyne *20a* was converted predominantly to *22a* with only a 20% contamination by the five-membered ring adduct *21a*. The unseparated mixture was directly converted to 2-methylarsabenzene *23a* in satisfactory yield.

Unfortunately, this method proved less satisfactory with more highly substituted 1,4-diynes. Stannohydration of 2,5-heptadiyne *20e* gave predominantly *21e*. The small quantities of 2,6-dimethylarsabenzene *23e* produced from mixtures of *21e* and *22e* were preparatively unsatisfactory. Even more discouraging was the observation that *20f* gave *21f* as the exclusive product of stannohydration. The formation of these five-membered ring adducts, useless for conversaion to heterobenzenes, can be explained by noting the greater stability of the intermediate radicals leading to *21f*.

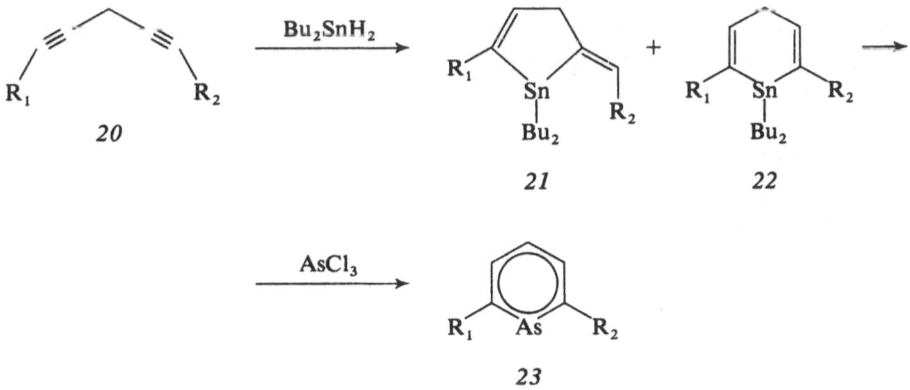

a, $R_1 = H$; $R_2 = CH_3$
b, $R_1 = H$; $R_2 = C_2H_5$
c, $R_1 = H$; $R_2 = t\text{-}C_4H_9$
d, $R_1 = H$; $R_2 = Si(CH_3)_3$
e, $R_1 = CH_3$; $R_2 = CH_3$
f, $R_1 = H$; $R_2 = C_6H_5$
g, $R_1 = H$; $R_2 = CH_2OAc$

On the other hand, this procedure has been used to prepare arsabenzenes bearing functional groups in the 2-position. Thus, 6-acetoxy-1,4-hexadiyne *20g* gave 2-acetoxy-methylarsabenzene *23g*, which on hydrolysis followed by oxidation gave 2-arsabenz-aldehyde *24*.

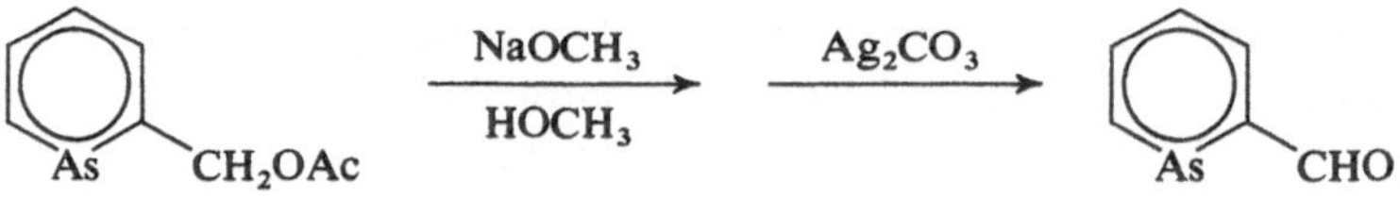

23g   *24*

Similarly, 3-substituted-1,4-pentadiynes *25* are converted to 4-substituted arsa-benzenes *27* [27, 28, 29]. 4-Acetoxyarsabenzene *27b* prepared in this manner was easily hydrolyzed to very useful 4-arsaphenol *28* [28].

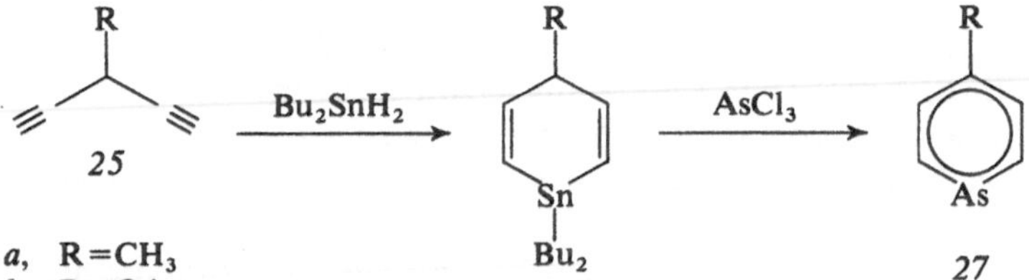

25

a, $R = CH_3$
b, $R = OAc$

26   *27*

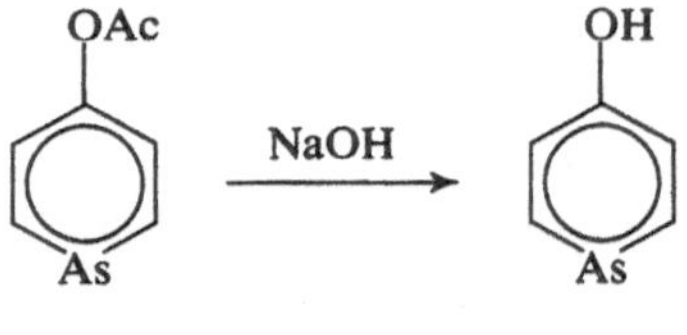

27b   *28*

4-Substituted heterobenzenes have also been obtained by direct functionalization of 1,1-dibutyl-1,4-dihydrostannabenzene *15* [30, 31].

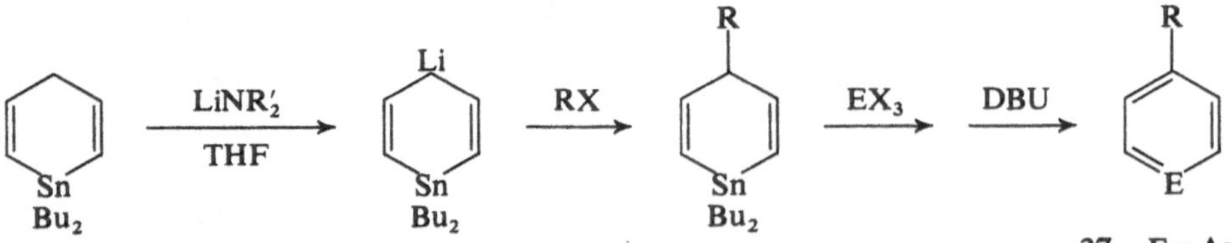

15   *26*

a, $R = CH_3$
b, $R = Si(CH_3)_3$
c, $R = t\text{-}C(CH_3)_3$

27, $E = As$
28, $E = Sb$
29, $E = Bi$

This adaption has proved quite valuable since it has been found that 4-substituted stibabenzenes *28* and bismabenzenes *29* are considerably more stable than the parent compounds [32, 33]. For example, *28a, b, c* are distillable liquids which only slowly polymerize at ambient temperature. Similarly, 4-t-butylbismabenzene *29c* is stable in dilute solution to 40 °C. These derivatives have allowed characterization which was not possible for the parent compounds.

Märkl has reported several valuable extensions of the dihydrostannabenzene route which have allowed the syntheses of a variety of 4-substituted arsabenzenes. 3-Substituted-3-methoxypentadiynes *30* were converted to the corresponding dihydrostannabenzenes *31*, which on treatment with arsenic trichloride gave 4-substituted arsabenzenes *24* [34, 35]. Intermediates *32c, d, e* lose the elements of methyl hypochlorite to form *24*.

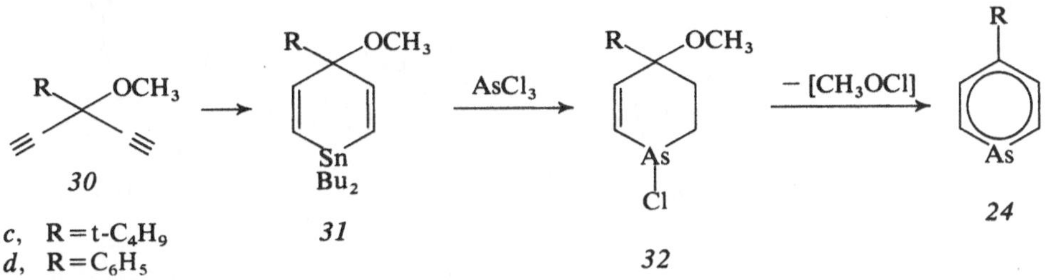

c, $R=t\text{-}C_4H_9$
d, $R=C_6H_5$
e, $R=C_6H_{11}$
f, $R=CH(OC_2H_5)_2$
g, $R=CO_2C_2H_5$

However, in the case of *32f*, the elements of diethoxy methyl chloride are lost to give 4-methoxyarsabenzene *33* [36]. Alternatively, if *32f* is reduced to secondary arsine *34*, methanol is eliminated to give *35* [37]. On hydrolysis, *35* affords 4-arsabenzaldehyde *36*. Similarly *32g* has been converted to *24g* which on hydrolysis gave 4-arsabenzoic acid *37* [38, 39].

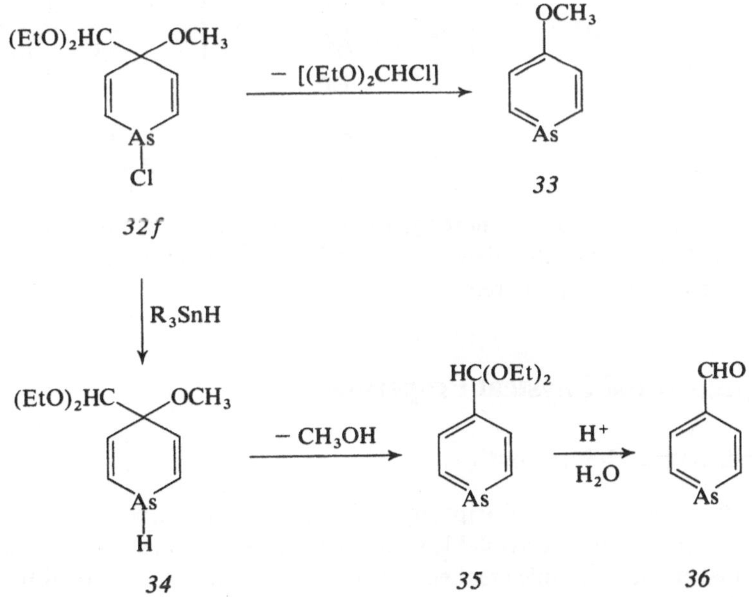

131

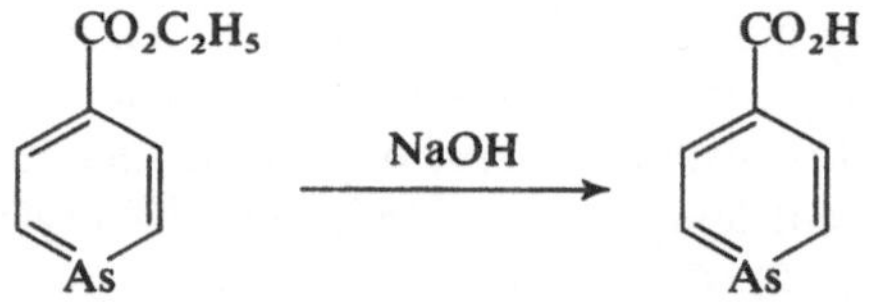

24 g   37

Finally, Märkl has developed two novel arsabenzene syntheses, which involve carbenoid ring expansion from arsacyclopentadienes. 1-t-Butyl-2,5-diphenylarsole *38* adds dichlorocarbene to give *39*, which on pyrolysis loses t-butyl chloride to afford 3-chloro-2,6-diphenylarsabenzene *40* [40].

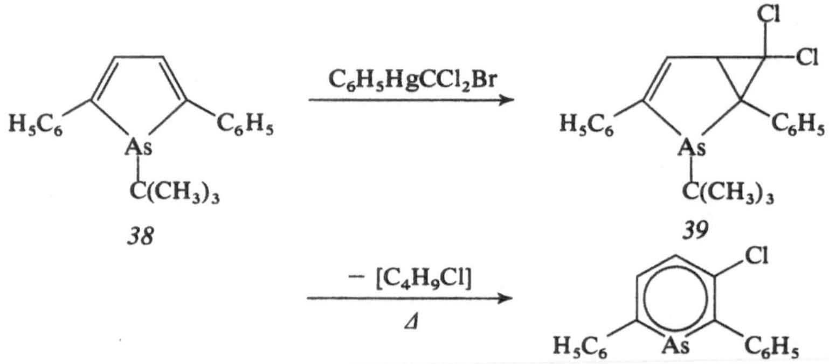

38   39

$- [C_4H_9Cl]$

Δ

40

In a related reaction, the pyrolysis of *41* gives *43*, apparently through carbene intermediate *42* [41, 42].

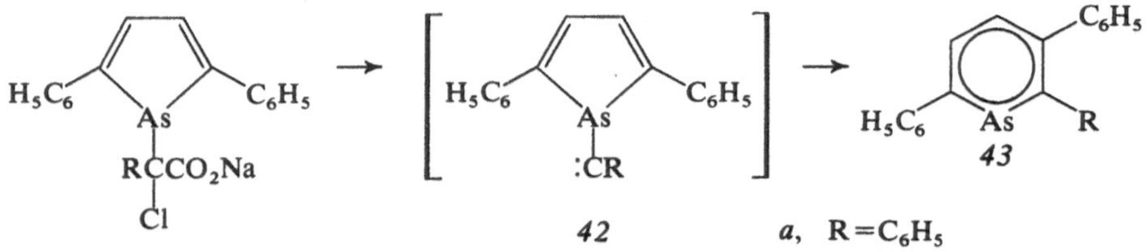

41   42   43

a, $R = C_6H_5$

b, $R = H$

The major limitation of these syntheses is that only 2,5-disubstituted or higher substituted arsacyclopentadienes are available [43]. Thus, only highly substituted arsabenzenes may be prepared.

3 Spectral and Physical Properties

3.1 Structure Determinations

A crystal structure of 2,3,6-triphenylarsabenzene *43a* has been determined [44]. Gas phase structural data are available from microwave spectral studies of arsabenzene [45] and stibabenzene [46], while arsabenzene has been studied by electron diffraction [47, 48].

The molecular structure of arsabenzene,[49] 4-methylarsabenzene[50] and 4-methyl-stibabenzene[50] have been determined from an NMR study of liquids oriented in a liquid crystal phase. Unfortunately, no structural data are yet available for bisma-benzene. Selected bond lengths and angles of the complete family of group 5 hetero-benzenes are illustrated in Fig. 1.

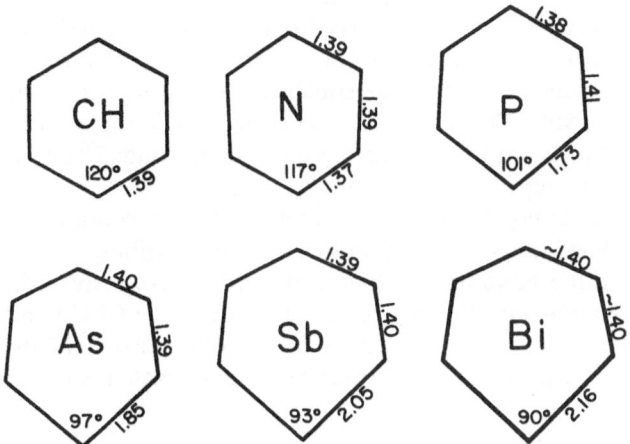

Fig. 1. Schematic Structures of the Molecules, C_5H_5E, where $E = CH$, N, P, As, Sb and Bi. The CEC bond angles and EC_α, $C_\alpha C_\beta$ and $C_\beta C_\gamma$ bond distances are indicated

All of the group 5 heterobenzenes are planar with C_{2v} point symmetry. The C-C bond lengths in the entire series vary only from 1.38 Å to 1.41 Å. This small bond alternation is less than is found in the benzocyclic compounds naphthalene and anthracene[51]. The average C-C bond distance of 1.395 Å is identical to that of benzene.

Table 1. A Comparison of the C-E bond Distances of the C_5H_5E Molecules with the Sum of the Covalent Radii of E and C

E	CE Bond Distance	Σ Covalent Radii[a]	Difference
CH	1.39	1.54	0.15
N	1.37[b]	1.50[c] (1.47)[d]	0.13 (0.10)
P	1.73	1.87	0.14
As	1.85	1.98	0.13
Sb	2.05	2.18	0.13
Bi	(2.16)[e]	2.29	(0.13)[e]

[a] Cotton, F. A., Wilkinson, G.: Advanced Organic Chemistry, 2nd ed., John Wiley and Sons, New York, 1966
[b] Sørensen, G. O., Mahler, L., Rastrup-Andersen, N.: *J. Mol. Struct.* 20, 119 (1974)
[c] Based on N_2H_4 for N
[d] Based on CH_3NH_2 for N
[e] Extrapolated value

The carbon-heteroatom (C-E) bonds of the heavier heterobenzenes are quite long in comparison to the C-C bond distances. See Table 1. Nevertheless, the C-E bonds are shorter than normal C-E single bonds. The E-P bond of phosphabenzene (1.73 Å) is 0.14 Å shorter than the sum of its covalent radii, while in both arsabenzene (1.85 Å) and stibabenzene (2.05 Å), the heteroatom-carbon bonds are 0.13 Å shorter. Extrapolation to bismabenzene suggests that the Bi-C bond distance is close to 2.16 Å. Obviously the C-E bonds of the heterobenzenes have multiple bond character.

Although planar, the heavier heterobenzenes are far from regular hexagons. The CEC bond angles show a progressive decrease from pyridine (117°) [52] through phosphabenzene (101°) and arsabenzene (97°) to stibabenzene suggests that the CBiC bond angle is close to 90°. The trend of decreasing bond angles with increasing atomic number is also shown by non-cyclic group 5 compounds [53] and is consistent with increasing p-character of the E-C bonds.

Of necessity, the ECC and CCC bond angles must exceed 120° to accommodate this change. However, this bond angle deformation is evenly distributed since there is little variation in the other bond angles. Indeed the progressive increase in the CE bond length in combination with the progressive decrease in the CEC bond angles serves to minimize the structural change in the carbocyclic portion of the C_5H_5E rings. Thus the very large heteroatoms can be accommodated without unusual strain.

3.2 NMR Spectra

The proton NMR spectra of the group 5 heterobenzens show a very characteristic AB_2X_2 pattern [54]. The two α-protons (X_2) lie far downfield from the β-proton multiplet (B_2) which in turn is downfield from the γ-proton signals (A). A typical spectrum, that of stibabenzene, is illustrated in Fig. 2, while the chemical shift values are tabulated in Table 2.

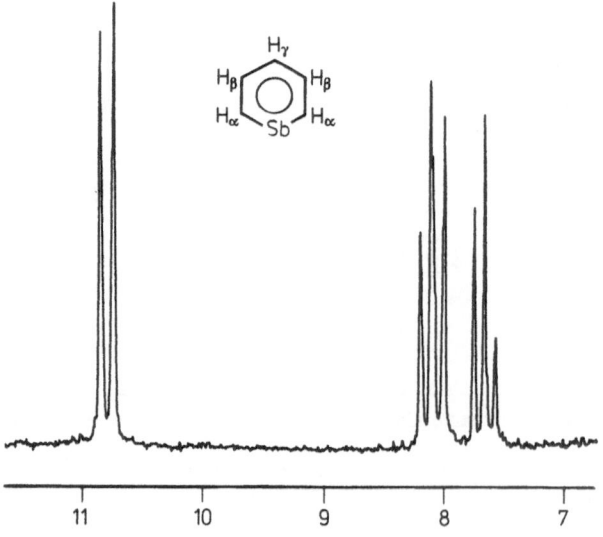

Fig. 2. The 100 MHz Proton NMR Spectrum of Stibabenzene

Table 2. Proton Chemical Shifts of Heterobenzenes[a]

Position	Benzene	Pyridine[b]	Phospha-benzene	Arsa-benzene	Stiba-benzene	Bisma-benzene
$H_\alpha(H_2, H_6)$	7.37	8.29	8.61	9.68	10.94	13.25
$H_\beta(H_3, H_5)$		7.39	7.72	7.83	8.24	9.8
$H_\gamma(H_4)$		7.75	7.38	7.52	7.78	7.8[c]

[a] Chemical shift values are in ppm downfield from Me_4Si
[b] Schneider, W. G., Bernstein, H. J., Pople, J. A.: Can. J. Chem. *35*, 1489 (1957)
[c] Assignment is only tentative due to relatively poor quality of the spectrum

All of the chemical shift values of the heavier heterobenzenes occur downfield from benzene. Particularly striking is the fact that the α-proton signals are at extraordinary low field and that this downfield shift increases sharply with atomic number of the group 5 atom. In the case of bismabenzene, the α-protons are nearly 6 ppm downfield from benzene. The β-proton signals show a similar but smaller progressive downfield shift, while even a very small shift is noted for the γ-proton signals.

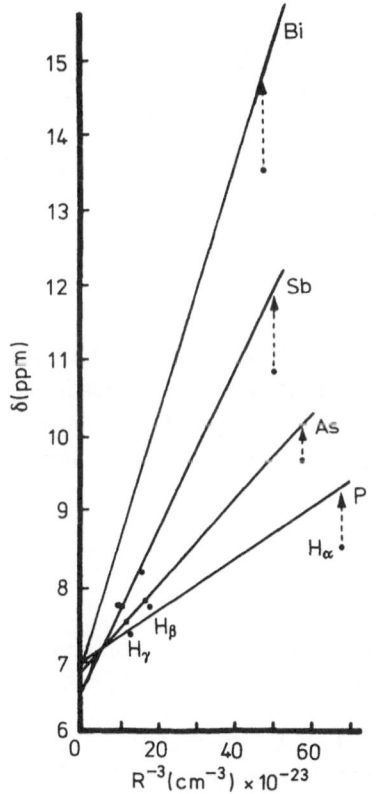

Fig. 3. The proton chemical shifts of the heterobenzenes vs. R^{-3}, where R is the interatomic distance between the group 5 atom and H_α, H_β or H_γ. The dashed lines indicate the estimated magnitude of the ortho-inductive effects. The slope of the lines indicates the magnitudes of the anisotropic effects of the group 5 atoms, while the intercept indicates the chemical shift value expected in the absence of this anisotropic effect

These large downfield shifts are readily explained in terms of the magnetic anisotropy of the group 5 atom [54]. Magnetic anisotropy is expected to increase with atomic number in agreement with the observed trend [55]. Furthermore, magnetic anisotropic effects on adjacent protons should be strongly distance dependent (R_{EH}^{-3} in the McConnell equation) [56] in agreement with the trend $\alpha > \beta > \gamma$. When the observed chemical shifts (after correction for electronegativity effects) [57] are plotted against the inverse cube of the distance between E and H (R_{EH}^{-3}), linear plots are obtained (Fig. 3). Extrapolation of R_{EH}^{-3} to 0 allows correction for the local anisotropic effect. After this correction, the chemical shift values are upfield from benzene. These non-local chemical shift values are consistent with smaller ring currents for the heterobenzenes.

The ^{13}C-NMR chemical shift data are collected in Table 3. The β- and γ-carbon shifts vary little from benzene to bismabenzene. However, the α-carbons are strongly deshielded and this deshielding is progressive in the series. For bismabenzene, the shift is nearly 90 ppm downfield from benzene. Qualitatively, the α-carbon shifts correlate with the α-proton shifts. However, carbon shifts of this magnitude are not easily explained by diamagnetic effects [54, 58].

Table 3. ^{13}C Chemical Shifts of Heterobenzenes and ^{1}J; ^{13}CH Values in Parentheses

	Benzene[a]	Phospha-benzene	Arsa-benzene	Stiba-benzene	Bisma-benzene[d]
δC_2	128.7 (159)[b]	154.1 (157)	167.7 (159)	178.3 (158)[c]	212 (—)[e]
δC_3		133.6 (156)	133.2 (157)	134.4 (153)[c]	133 (150)
δC_4		128.8 (161)	128.2 (161)	127.4 (—)	—

[a] Stothers, J. B.: Carbon-13 NMR Spectroscopy, Academic Press, New York, N.Y., 1972
[b] J values in parentheses; hertz
[c] The $^1J_{^{13}CH}$ values recorded are those for the more stable 4-methylstibabenzene
[d] Recorded for the more stable 4-t-butylbismabenzene
[e] The signal for C_2 was too broad to measure the J_{CH} value

The $^1J_{CH}$-values have been measured for all the heterobenzenes. All values are extremely close to that of benzene. Indeed, the larger variation for the 4-substituted stiba- and bismabenzenes may be a substituent effect. $^1J_{CH}$ values are a sensitive function of hybridization [59]. Thus, the near identity of all the $^1J_{CH}$ values indicates a hybridization identical to that of benzene.

3.3 Mass Spectra

All of the group 5 heterobenzenes show mass spectra typical of aromatic compounds (Table 4). The compounds show intense molecular ions which are the base peaks for all the heterobenzenes except bismabenzene. Important fragmentation involves loss of either C_2H_2 or HCE from the molecular ion. Loss of HCE becomes relatively less important for the heavier heterobenzenes. This observation may be explicable in terms of declining stability of the HCE molecules. However, the heavier hetero-benzenes show increasing peaks for E^+ and for loss of E. This tendency is consistent with weaker E-C bonding and with increasing positive character of elements E.

Table 4. Mass Spectra of Benzene and the Heterobenzenes C_5H_5E, $E = N$, P, As, Sb and Bi, Showing the Major Ions and their Relative Abundances

Compound	Parent Ion $(M)^+$	$(M-C_2H_2)^+$	$(M-HCE)^+$	$(M-E)^+$	E^+
$C_6H_6{}^a$	78 (100)	52 (18)	—	—	—
$C_5H_5N^b$	79 (100)	53 (8)	52 (6)	—	—
$C_5H_5P^b$	96 (100)	70 (24)	52 (8)	65 (0)	—
$C_5H_5As^b$	104 (100)	114 (24)	52 (1)	65 (1)	75 (2.5)
$C_5H_5Sb^b$	{ 188 (65) / 186 (100)	{ 162 (9) / 160 (16)	52 (0)	65 (95)	{ 123 (18) / 121 (27)
$C_5H_5Bi^c$	274 (30)	252 (0)	52 (?)^c	65 (?)^c	209 (100)

^{a} Cornu, A., Massot, R.: Compilation of Mass Spectral Data, Heyden, London, 1974, v. 1, p. 11A
^{b} The mass spectra were determined on an AEl-MS902 spectrometer at 45 eV using a gas inlet
^{c} Low mass peaks were obscured by impurities

3.4 Photoelectron (PE) and Electron Transmission (ET) Spectra

The UV-PE spectra of all of the heterobenzenes have been determined [60-62] while ET spectra are available for all except bismabenzene [63]. PE spectroscopy allows observation of various mono-ionization potentials of molecules which correspond to differences in energy between the ground state and different radical cation states [64]. Within the context of Koopmans' theorem, these ionization potentials

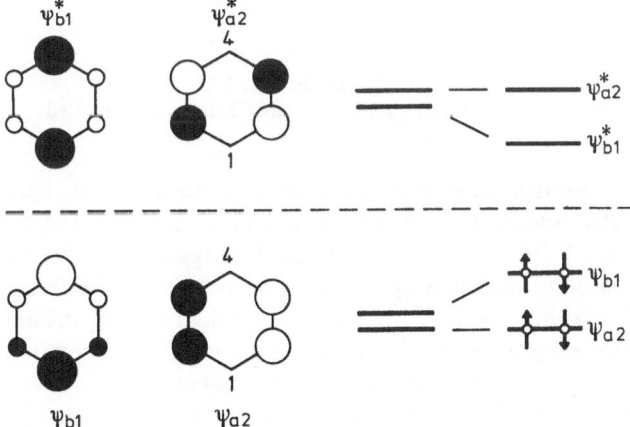

Fig. 4. Diagrams of the degenerate HOMO (ψ_{b1} and ψ_{a2}) and LUMO (ψ_{b1}^* and ψ_{a2}^*) of benzene as split in the C_{2v} symmetry of the heterobenzenes, C_5H_5E. The shaded and non-shaded circles indicate different signs of the HMO coefficients, while the areas are proportional to the squares of the coefficients. Atom E is located at center 1

$(I_j = -\varepsilon_j)$ are associated with the bonding molecular orbital energies of the ground state. In a similar manner, ET spectroscopy allows measurements of various electron affinities which correspond to differences in energy between the ground state and different radical anion states [65]. These electron affinities are then associated with antibonding orbital energies of the ground state. In concert, these techniques can be used to deduce the nature of the frontier molecular orbitals of molecules.

In view of their similarity the heterobenzenes may be treated as perturbed benzenes. Figure 4 illustrates the highest occupied and lowest unoccupied molecular orbitals (HOMO and LUMO) of benzene. In the C_{2v} symmetry of the heterobenzenes these orbitals split into non-degenerate a_2- and b_1-orbitals. To the first approximation replacement of C_1 by a heteroatom E should leave the a_2-orbitals unchanged since they have a node at this position. However, the b_1-orbitals which have an antinode at the heteroatom should be strongly effected.

Experimentally the PE spectra of C_5H_5P, C_5H_5As, C_5H_5Sb and C_5H_5Bi show the first ionization potential to be strongly shifted to lower energy with increasing atomic number of the heteroatom (see Table 5). The overlapping bands for the second and third ionization potentials show only small shifts. On this basis, the b_1-orbital must be the HOMO, while the next bands correspond to ionization from the $\pi(a_2)$ and $n(a_1)$ orbitals. These $\pi(a_2)$ and $n(a_1)$ bands also have very similar energy in pyridine, but the $\pi(b_1)$ band is of higher energy. Thus the ordering of π-M.O.'s is inverted between pyridine and the heavier heterobenzenes.

Table 5. Comparison of $\pi(b_1)$ Ionization Potentials of Molecules C_5H_5E with the Ionization Potential of Free Atom E^a and the C-E Bond Length of $C_5H_5E^b$

E	b_1-π I.P.	E.I.P.	CE Bond Distance
N	10.5 ev	14.53 ev	1.37 Å
P	9.2	11.0	1.73
As	8.8	9.81	1.85
Sb	8.3	8.64	2.05
Bi	7.9	7.29	(2.16)

[a] Bastide, J., Heilbronner, E., Maier, J.-P., Ashe, A. J., III: Tetrahedron Lett. *1976*, 411
[b] Burrow, P. D., Ashe, A. J., III, Bellville, D. J., Jordan, K. D.: J. Am. Chem. Soc. submitted

The large shift in the b_1-orbitals with increasing atomic number has been associated with decreasing electronegativity of the heteroatoms as measured by free atom ionization potentials [60,61]. Alternatively, it can be argued that the decreasing strength of the C-E π bond, which is appropriate for increasing C-E bond distances, will increase the b_1-orbital energy [63,66]. It can be noted that b_1-orbital ionization potentials correlate directly with the free atom ionization potentials and inversely with the measured C-E bond lengths of the heterobenzenes.

Results from ETS indicate that the LUMO's of the heterobenzenes are strongly stabilized and thus have b_1-symmetry [63]. This trend is consistent with a decrease in E-C π-antibonding which should parallel the decrease in π-bonding.

Molecular orbital calculations on phosphabenzene reproduce this ordering of frontier M.O.'s [67,68]. Although arsabenzene has been subjected to less theoretical work, *ab initio* [69], CNDO/2 [70] and CNDO/S [71] calculations indicate that the HOMO has b_1-symmetry. Unfortunately, stibabenzene and bismabenzene have not been treated beyond an empirical HMO parameterization from their PE spectra [72,73].

The ordering of the HOMO and LUMO's provides a satisfying explanation for the enhanced stability of the 4-substituted heterobenzenes and the 10-substituted hetero-anthracenes. In the FMO treatment an attacking reagent must be able to effectively mix its HOMO and LUMO with those of the heterobenzenes [74]. Because the C_4-position has the largest orbital coefficient in both the HOMO and LUMO, it is predicted to be the site of attack. Thus, steric blocking of the most reactive position greatly increases the kinetic stability. It is clear from the PE spectra of 9-heteroanthracenes that the 10-substituents have only minor electronic effects [75].

3.5 Dipole Moments

The gas-phase dipole moment of arsabenzene was found to be 1.10D [45]. In cyclo-hexane solution it was measured as 1.02D [27]. These values are typical of those found for tertiary arsines: for trimethylarsine $\mu = 0.86$D [76], for triethylarsine $\mu = 1.04$D [77] and for triphenylarsine $\mu = 1.23$D [78]. No dipole moment data are available for stibabenzene and bismabenzene. However, based on trends shown in acyclic compounds [79], it is expected that they have smaller moments.

M.O. calculations have predicted that all of the heterobenzenes have the same direction of polarization with negative end of the dipole towards the hetero-atom [67-70]. This has been confirmed by substituent effects from pyridine, phos-phabenzene and arsabenzene [27]. In each case the 4-methyl derivatives have dipole moments which exceed the parent compound since the electron donating methyl group reinforces the ring dipole.

4 Chemical Properties

4.1 Diels-Alder Reactions

Benzene and all of the group 5 heterobenzenes except pyridine behave as dienes in Diels-Alder reactions [11]. Qualitatively reactivity increases with the atomic number of the heteroatom. For example, benzene reacts with hexafluorobutyne at 180° [80], while phosphabenzene gives adduct *44b* at 100°. Arsabenzene reacts at 25° while stibabenzene and bismabenzene react at 0°.

44a, E=CH
  b, E=P
  c, E=As
  d, E=Sb
  e, E=Bi

Diels-Alder reactivity of benzocyclic aromatics towards a common dienophile has been correlated with the loss of resonance energies as measured by their 1,4-localization energies [81]. In the case of the heterobenzenes, the relationship must be more complicated even assuming the same mechanism. Since different E-C single bonds are formed in 44, differential reactivity will depend on the difference between loss or resonance energies of the heterobenzenes and the gain in energy from forming E-C single bonds of 44. Since E-C single bond strength is known to decrease in the group 5 column [82], the resonance energies must decrease.

At low temperature stibabenzene and bismabenzene reversibly form Diels-Alder dimers 45 [24,32,33]. The greater delocalization energies of phosphabenzene and arsabenzene must prevent them from forming dimers. Since the dimerizations are equilibria, it is likely that dimers 46 are higher in energy than 45. Presumably the greater strength of a C-C bond and an E-E bond over two E-C bonds explains this difference [83]. Interestingly, 1-methylsilabenzene forms dimer 47 [84,85]. However, 47 is a kinetic product since its formation is irreversible.

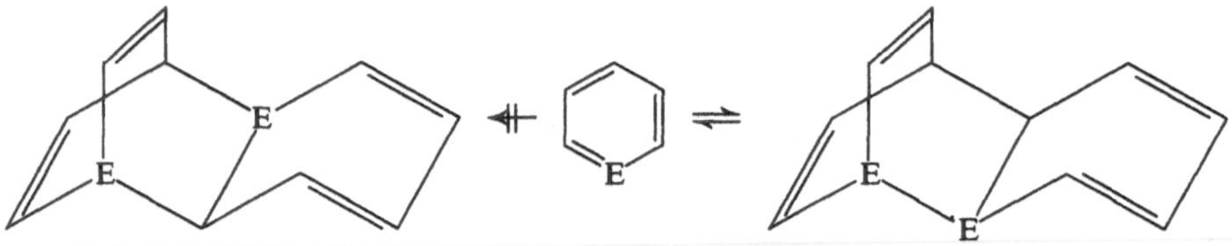

46
a, E=Sb
b, E=Bi
45

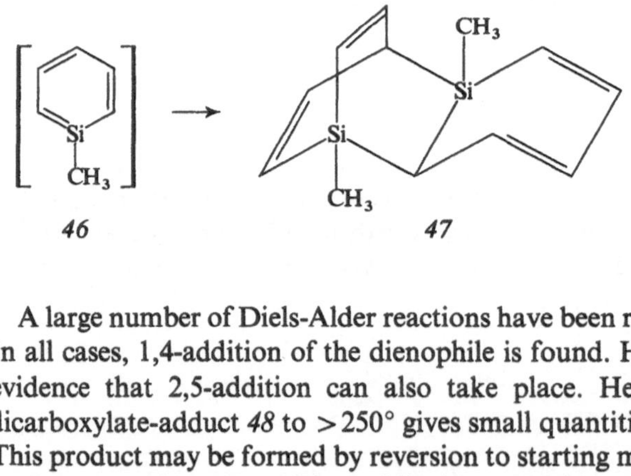

46   47

A large number of Diels-Alder reactions have been reported for arsabenzenes [86,87]. In all cases, 1,4-addition of the dienophile is found. However, there is some indirect evidence that 2,5-addition can also take place. Heating the dimethyl acetylene-dicarboxylate-adduct 48 to >250° gives small quantities of dimethyl phthalate 49 [87]. This product may be formed by reversion to starting materials followed by readdition to form transient 50 which rapidly loses HCAs to give 49. However, the major product of pyrolysis of 48 is Alder-Richart cleavage to substituted arsabenzene 51.

A variation of this procedure provides a mild and preparatively useful route to 2- and 3-substituted arsabenzenes [88,89]. Tetrazine 52 reacts with 48 to give good yields of 51 along with the expected diazine 53. Application of this method has allowed the syntheses of 1-arsanaphthalene 55 [90]. Arsanaphthalene is rather labile but may be readily characterized by conversion to adduct 56.

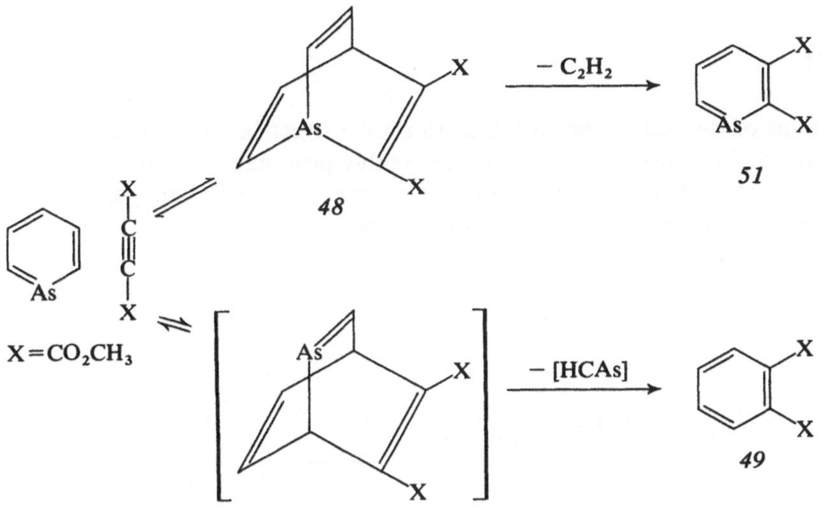

X = CO$_2$CH$_3$

$-$ C$_2$H$_2$

$-$ [HCAs]

48

49

50

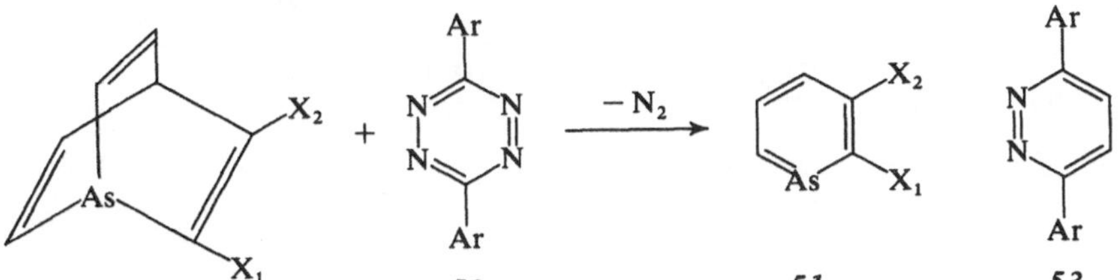

51

48 + 52 $\xrightarrow{- N_2}$ 51 53

a, X$_1$ = X$_2$ = COCH$_3$
b, X$_1$ = H,   X$_2$ = CO$_2$CH$_3$
c, X$_1$ = CO$_2$CH$_3$, X$_2$ = H
d, X$_1$ = COCH$_3$,   X$_2$ = H
e, X$_1$ = H,   X$_2$ = COCH$_3$

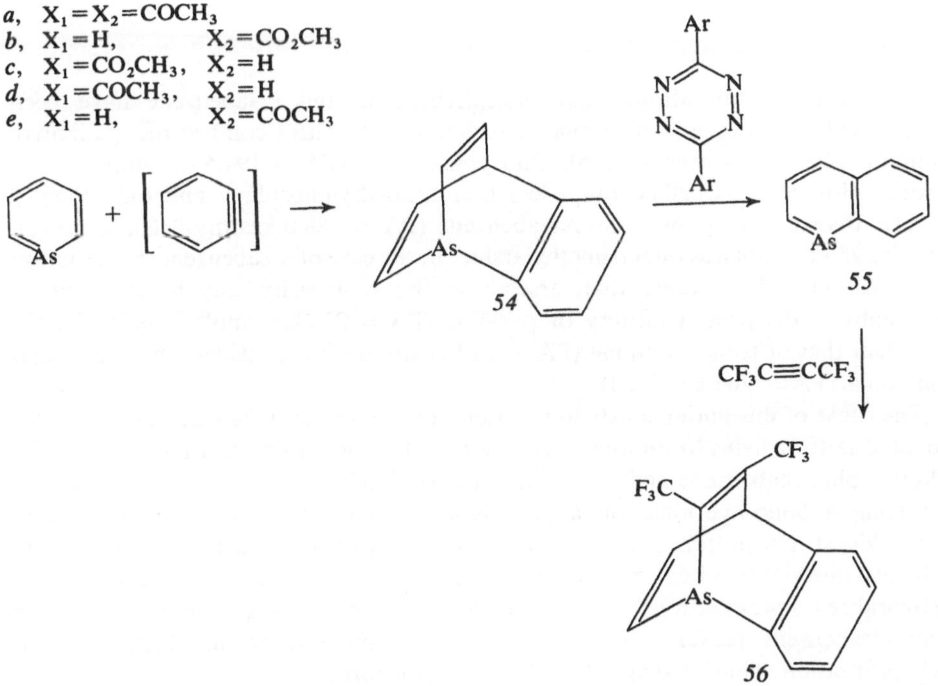

54

55

CF$_3$C≡CCF$_3$

56

141

4.2 Basicity

The heteroatoms of the heavier heterobenzenes show neither basic nor nucleophilic properties. For example, arsabenzene is not detectably protonated in trifluoroacetic acid [91], nor can it be alkylated with methyl iodide or stronger alkylating agents [92]. This contrasts with simple tertiary arsines which show modest basicity [93] and which are good nucleophiles:

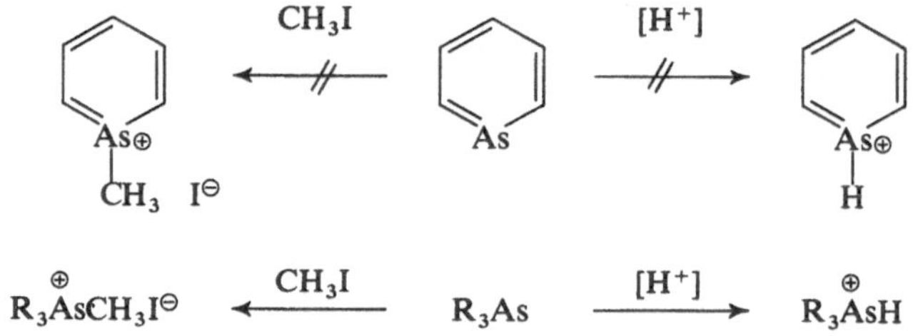

$$\text{As} \xleftarrow{\;\;CH_3I\;\;//\;} \text{As} \xrightarrow{\;\;[H^+]\;\;//\;} \text{As}$$

As⊕CH₃ I⊖ (left structure) As (center) As⊕H (right structure)

$$R_3\overset{\oplus}{As}CH_3I^\ominus \xleftarrow{\;CH_3I\;} R_3As \xrightarrow{\;[H^+]\;} R_3\overset{\oplus}{As}H$$

Gas phase proton affinities of phosphabenzene and arsabenzene have been determined by ion-cyclotron resonance techniques [94]. These confirm the qualitative solution phase data (see Fig. 5). Phosphabenzene (PA = 194.5 kcal/mol) has a proton affinity nearly 30 kcal/mol less than trimethylphosphine and only slightly greater than that of phosphine. Arsabenzene (PA = 188.0 kcal/mol) has a proton affinity 23 kcal/mol less than trimethylarsine. In the case of arsabenzene, protonation occurred on carbon rather than arsenic so the As-basicity may be even lower. By contrast, the proton affinity of pyridine (PA = 218 kcal/mol) is only slightly less than that of trimethylamine (PA = 222 kcal/mol) but considerably larger than ammonia (PA = 202 kcal/mol).

The cause of this anomalously low basicity of the heavier heterobenzenes appears to be due to a hybridization-geometric effect. Organophosphines and arsines (including phosphabenzene and arsabenzene) have CEC bond angles close to 100° [95], implying a bond-hybridization approaching p³. The lone pair must be largely s-hybridized. On protonation, the lone pair must gain p-character to take part in effective directed bonding. Acyclic phosphines and arsines are geometrically able to rehybridize on protonation [96]. However, the rigidity of the rings of phosphabenzene and arsabenzene prevent charges in the CEC bond angles and hence prevent rehybridization. Thus, a strong E-H bond cannot form.

In the case of the nitrogen compounds, the bond angles in the unprotonated species are rather close to those of the protonated species [97]. Since there are only rather small geometric changes, the proton affinity of pyridine is close to that of trimethylamine.

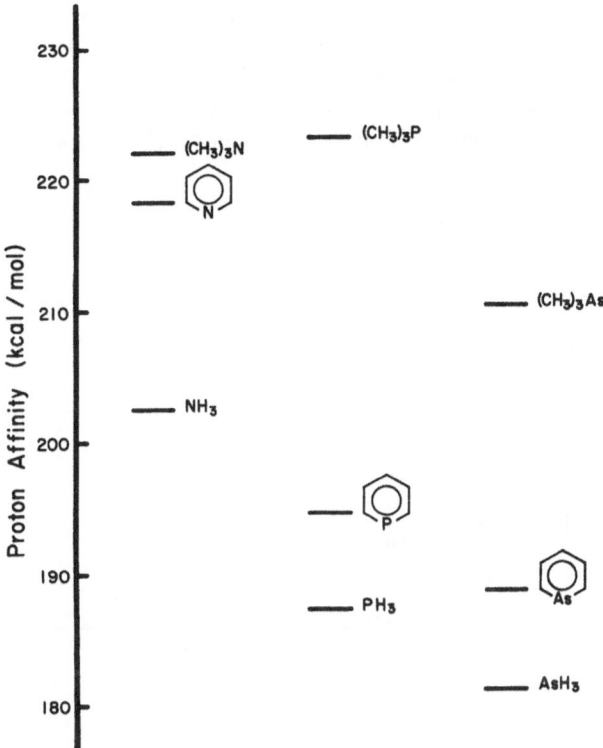

Fig. 5. Gas phase proton affinities of pyridine, phosphabenzene and arsabenzene compared to selected amines, phosphines and arsines

Results from X-ray PE spectra studies are consistent with this analysis [98]. Core ionization energy of the arsenic 3s electrons of arabenzene (211.2 ev) was almost identical to that of trimethylarsine (211.1 ev) but considerably less than that of arsine (212.4 ev). Core ionization of the phosphorus 2p electrons of the phosphorus series followed a similar trend. Ionization of core electrons is a vertical process which is sensitive to the ability of an atom to bear positive charge without geometric change. On the other hand, proton affinity is adiabatic and depends not only on the ability to bear charge but also on the ability to undergo stabilizing geometric change on protonation. In fact there are a number of excellent correlations between proton affinity and core ionization for several series of compounds which undergo similar geometric changes on protonation [99]. The failure of the proton affinity-core ionization correlation for arsabenzene and trimethylarsine suggests the compounds have different geometric requirements for protonation.

4.3 Transition Metal Complexation

While arsabenzene does not act as a nucleophile toward "hard" acids, it does form σ-Mo(CO)$_5$ complex *57* on treatment with pyridine-Mo(CO)$_5$ and boron trifluoride etherate [100]. Qualitatively complex *57* seems rather weak since on heating it is destroyed, forming small quantities of π-Mo(CO)$_3$complex *58* [101]. This π-complex is more conveniently prepared directly from arsabenzene and Mo(CO)$_6$ or from acid-catalyzed displacement from tris-(pyridine)molybdenum tricarbonyl.

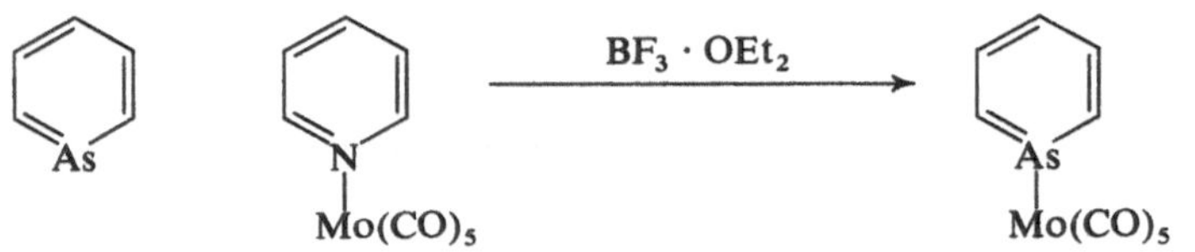

Stibabenzene-Mo(CO)$_3$ *59* may be prepared by the latter method. However, bismabenzene is destroyed by BF$_3$ · OEt$_2$. The π-basicity of the heterobenzenes appears to be comparable to that shown by benzocyclic aromatics.

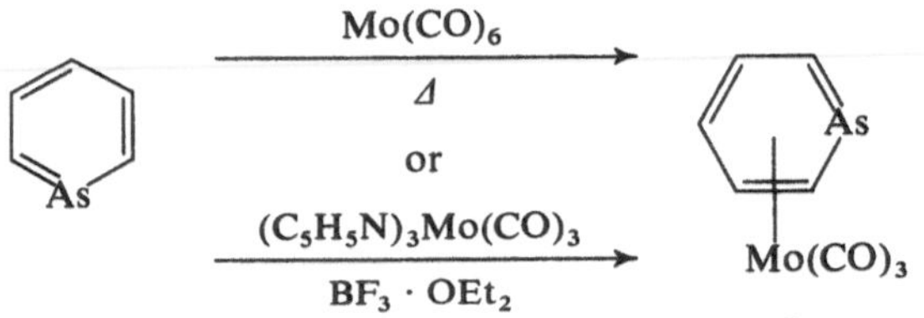

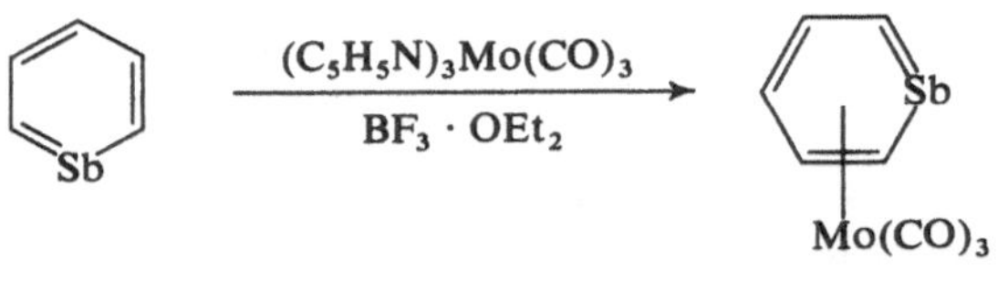

4.4 Electrophilic Aromatic Substitution

Although arsabenzene is more sensitive to oxidation than are normal benzocyclics, it undergoes a variety of electrophilic substitution reactions [87,88]. Substitution takes place at the 2- and 4-positions; in no case have 3-substituted products been detected.

144

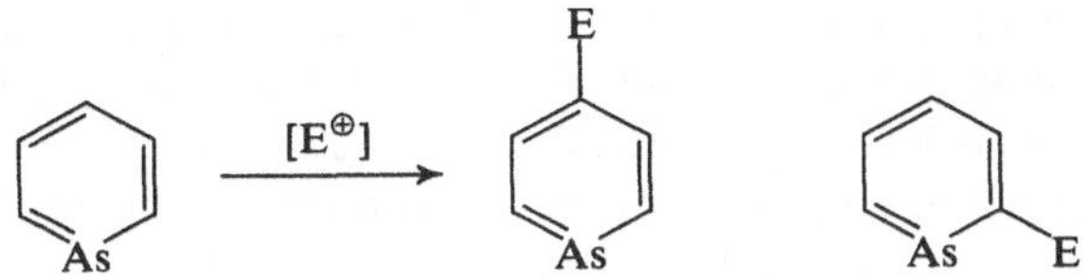

60   61

a, E=COCH₃
b, E=NO₂
c, E=D

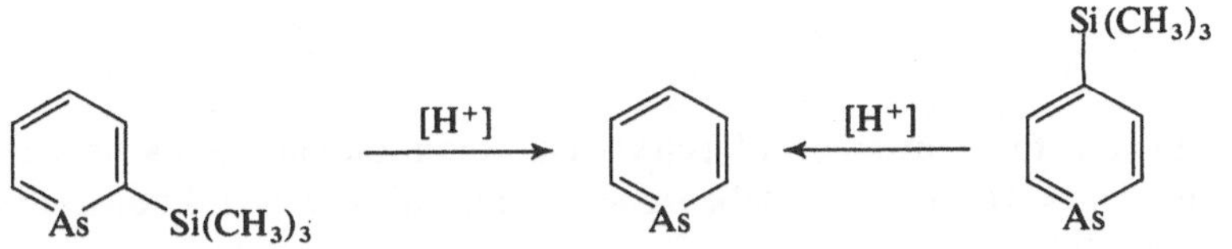

62   63

Friedel-Crafts acetylation of arsabenzene at −78° in methylene chloride using acetyl chloride-aluminium chloride gave an 86% yield of a 4:1 mixture of 4-acetyl-arsabenzene *60a* and 2-acetylarsabenzene *61a*. Nitration at 0° with nitric acid in acetic anhydride gave only a 20% yield of a 2:1 mixture of 4-nitroarsabenzene *60b* and 2-nitroarsabenzene *61b*. Deuterium exchange using trifluoroacetic-d₁ acid takes place in the 2- and more slowly in the 4-position. Protodesilylation of 2- and 4-trimethylsilylarsabenzenes (*62* and *63*) takes place readily in trifluoroacetic acid. Unfortunately, 3-trimethylsilylarsabenzene is unavailable for comparison. The conditions necessary for halogenation and sulfonation destroy arsabenzene.

Competition experiments with benzocyclic aromatics show that arsabenzene is considerably more reactive than benzene. Arsabenzene is acetylated at approximately the same rate as mesitylene, that is about 10³ faster than benzene. Deuterium exchange takes place at a rate comparable to that of p-xylene, again about 10³ faster than benzene.

Qualitatively, the effect of the electropositive heteroatom of arsabenzene appears to be comparable to that of an activating ortho-para directing substituent on benzene. Presumably, the electrophile attacks only at the 4- (and 2-)positions because only these positions allow efficient electronic stabilization of the intermediate σ-complex *64*.

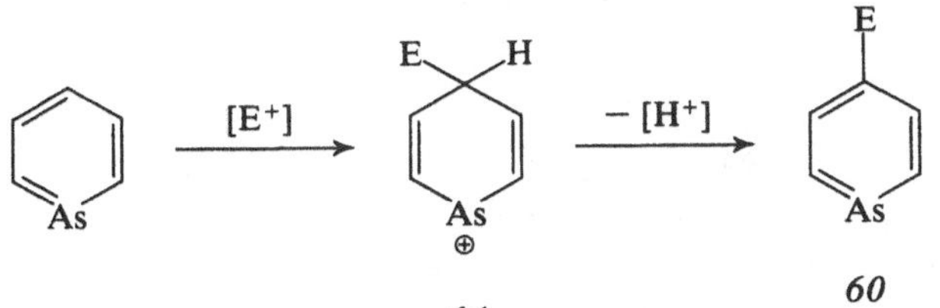

64   60

145

Several electrophilic rearrangements of arsacyclohexadienes appear closely related to the direct substitution reaction. 4-Substituted-1-aryl-4-methoxycyclohexadienes *65* are smoothly converted to 4-substituted-2-aryl-arsabenzenes *68* on treatment with acid [102]. Presumed intermediate *67* is, of course, analogous to the intermediates in the direct electrophilic substitution.

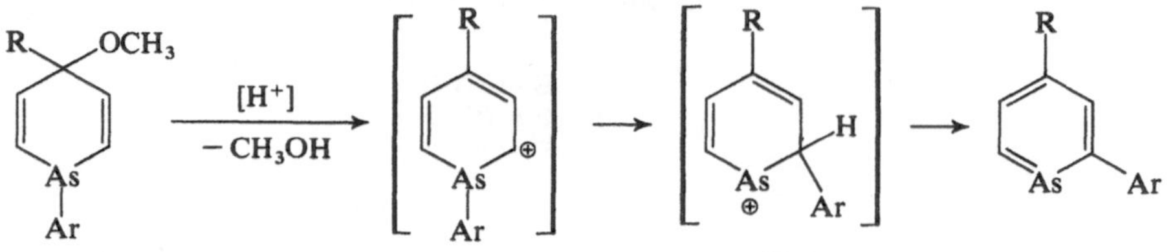

65 *66* *67* *68*

Similarly, the conversion of 1-aryl-arsacyclohexadienones *69* to 2-aryl-4-acetoxy-arsabenzenes *70* is the arsabenzene analog of the well-known dienone-phenol rearrangement [103].

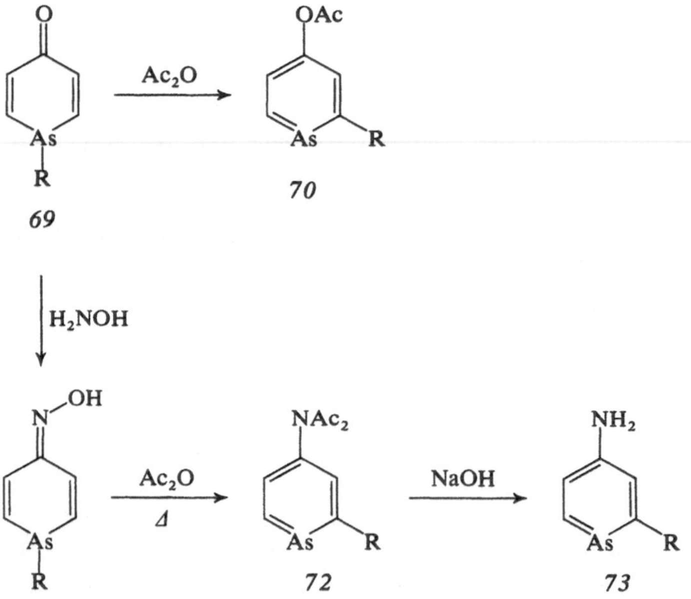

69 *70* *71* *72* *73*

a, R=Ar
b, $R=CH_2C_6H_5$
c, R=H

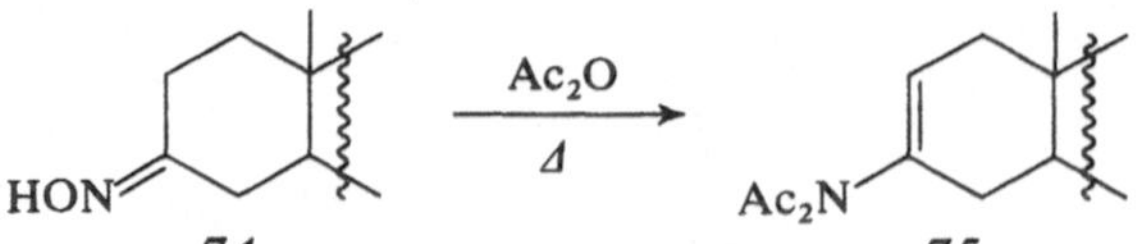

74 *75*

The conversion of oxime *71* to imide *72* is a related but more complicated reaction [104]. The reduction of the oxime concurrent with migration finds analogy in the conversion of steroid ketoximes *74* to enimides *75* under identical conditions [105]. The imide *72* may be hydrolyzed to arsaniline *73*. It might also be noted that in the case of benzyl substituents *69b* and *71b* the reactions take place with partial loss of substituent (to *70c* and *72c*). Thus, these procedures may be used to prepare the unsubstituted 4-arsaniline and 4-arsaphenol.

4.5 Free Radical Rearrangements

Although mechanistic studies are lacking, several aromatic-type rearrangements have been reported which appear to involve arsabenzyl radicals. Thus, the thermal rearrangement of 4-methylene arsacyclohexadiene *76* to *77* appears to be analogous to the von Auwers semibenzene-benzene rearrangement [106]. Very likely a radical dissociation-recombination is involved.

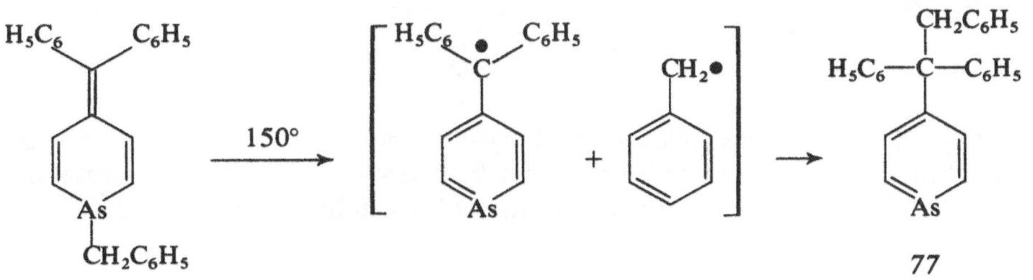

More speculative is the suggestion that the reaction of *78* with AsCl$_3$, which gives 46% of *79* on strong heating, involves arsabenzyl radical coupling [106].

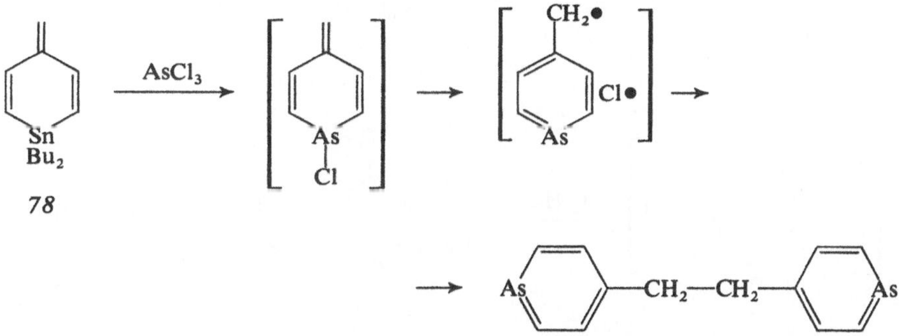

Chan has found that the reaction of 2-arsabenzenemethanol (*80*) with triethyl-orthoacetate gave *82* and *83* [107]. Probable intermediate *81* [108] may rearrange to *82* by addition-elimination of arsabenzyl radicals, which combine to give *83*. This reaction appears similar to the well-studied conversion of α-benzyloxystyrene *84* to β-phenylpropiophenone *85* [109].

147

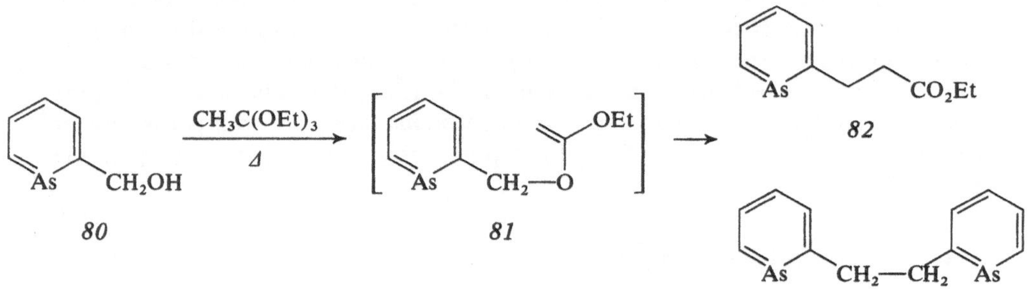

Finally, Bickelhaupt has found that 10-aryl-9-arsaanthracene *6b* is formed from the pyrolysis of 9-aryl-10-benzyl-9,10-dihydro-9-arsaanthracenes *86* [110]. The reaction has been shown to occur by homolysis to radicals, followed by a 1,4-aryl shift and hydrogen abstraction [111].

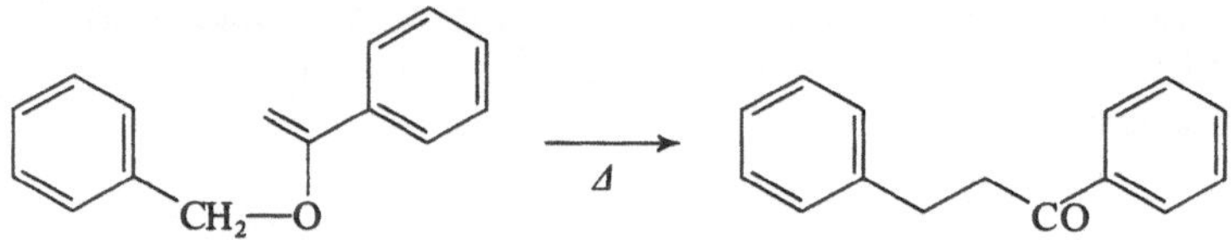

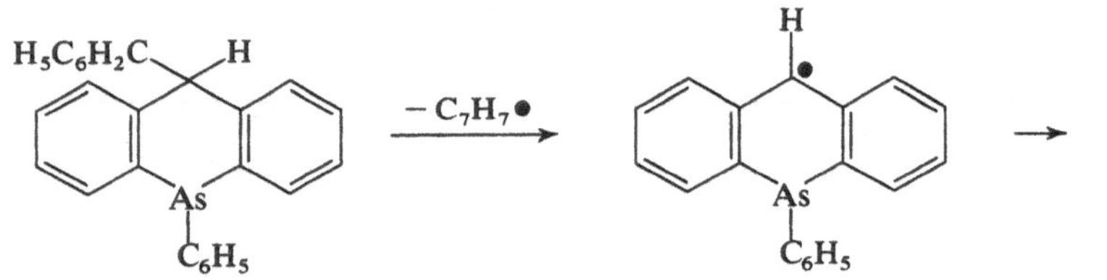

4.6 Reactions with Alkyllithium Reagents

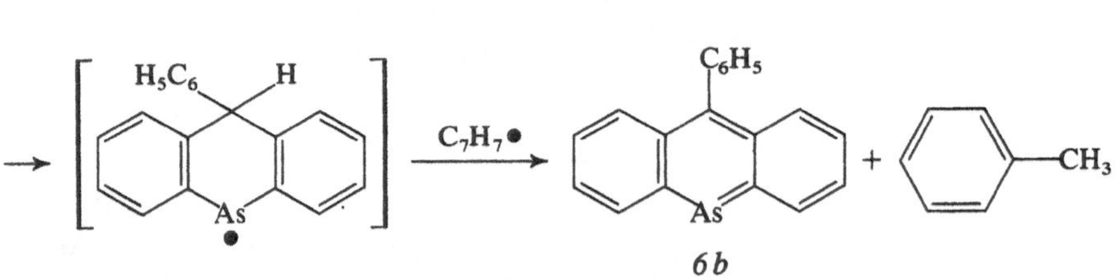

Pyridine is attacked by alkyllithium reagents at its most electropositive atom (C_2) to give the charge delocalized anion *87* [112]. In a similar manner the heavier hetero-benzenes are attacked at their most electropositive atoms (the heteroatoms) to give heterocyclohexadienides *88* [113].

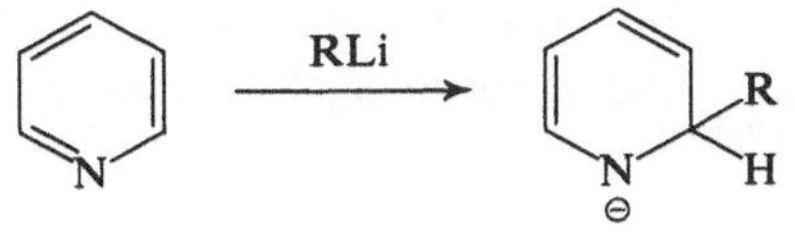

87

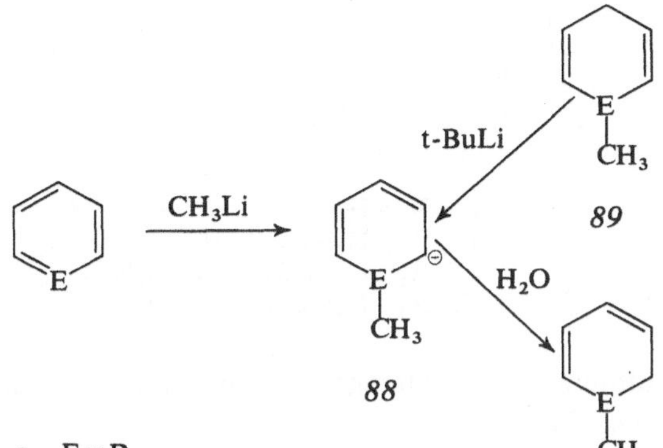

a, E=P
b, E=As
c, E=Sb

For example, methyllithium in ether-THF adds to arsabenzene to give a dark-green solution of anion *88b* which can be generated independently by treating 1-methyl-2,5-arsacyclohexadiene *89b* with t-butyllithium. Quenching with water affords *90b*.

1-Methylarsacyclohexa-2,4-diene *90b* has been converted to a λ^5-arsabenzene [114]. Quaternarization of *90b* with methyl iodide followed by deprotonation gave 1,1-dimethyl-λ^5-arsabenzene *92*.

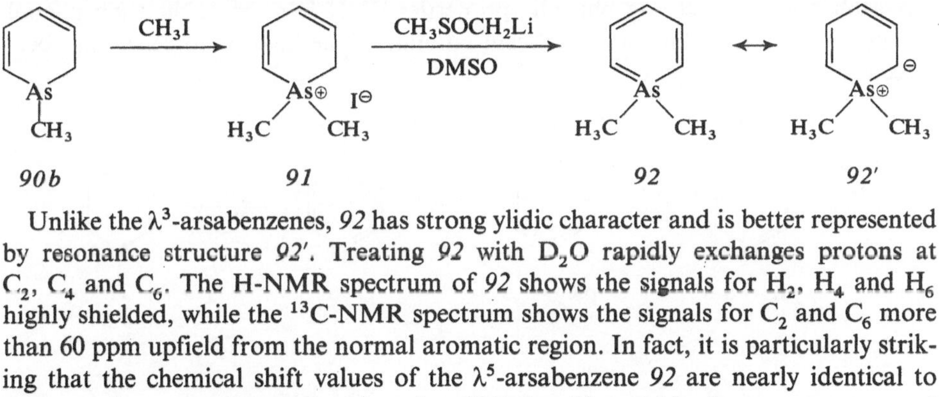

90b   *91*   *92*   *92'*

Unlike the λ^3-arsabenzenes, *92* has strong ylidic character and is better represented by resonance structure *92'*. Treating *92* with D_2O rapidly exchanges protons at C_2, C_4 and C_6. The H-NMR spectrum of *92* shows the signals for H_2, H_4 and H_6 highly shielded, while the ^{13}C-NMR spectrum shows the signals for C_2 and C_6 more than 60 ppm upfield from the normal aromatic region. In fact, it is particularly striking that the chemical shift values of the λ^5-arsabenzene *92* are nearly identical to those of the arsacyclohexadienide anion *88b*. On this solid basis, we cannot regard the λ^5-arsabenzenes as aromatic.

4.7 Miscellaneous Functional Group Chemistry

Although a variety of different functionally substituted arsabenzenes have now been synthesized, to date only 4-arsabenzaldehyde, 4-arsaphenol and the arsabenzoic acids have been extensively studied.

Arsabenzaldehydes appear to be normal aromatic aldehydes [26,37,115]. 4-Arsabenzaldehyde (*36*) undergoes the Aldol and Knoevenagel as well as a variety of typical aldehyde addition reactions [37,115].

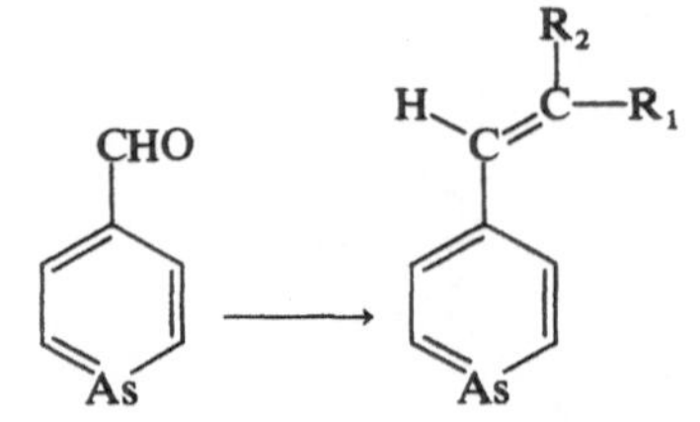

36   93   a, R₁=H,   R₂=COCH₃
                   b, R₁=R₂=CO₂C₂H₅
                   c, R₁=CO₂C₂H₅, R₂=CN
                   d, R₁=R₂=CN

a, R₁=H, R₂=COCH₃

However, with phenyllithium or phenylmagnesium bromide addition to arsenic takes preference to aldehyde addition [115].

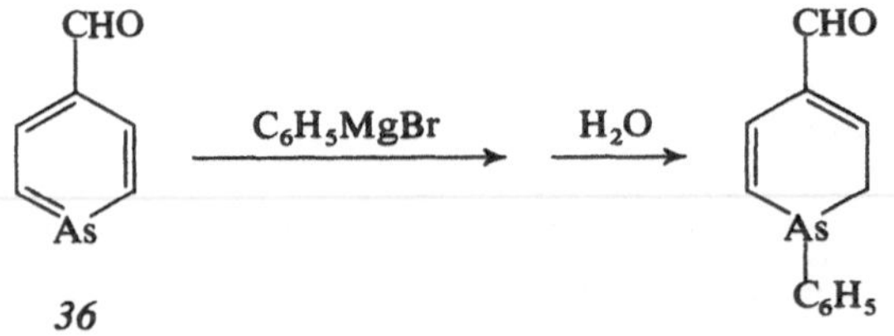

36

94

Arsaphenol shows strong phenolic properties [25]. It is more acidic than phenol. Unlike 4-hydroxypyridine, no significant amount of the keto tautomer *95* has been detected.

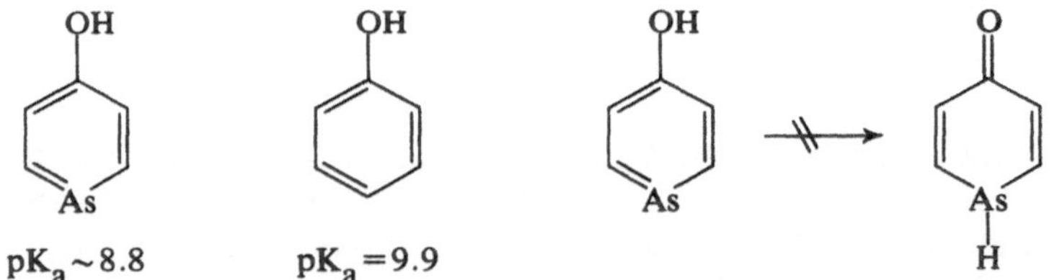

$pK_a \sim 8.8$ $pK_a = 9.9$

95

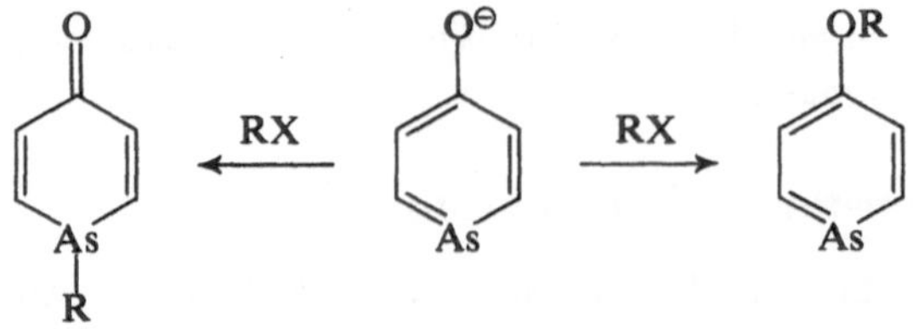

96

97

Arsaphenoxide shows ambient properties [116]. Reaction with organic halides gives either O- or As-alkylation depending on conditions and the alkylating agent. Particularly interesting was the reaction of 4-arsaphenol with allylbromide to give triallyl 4-arsacyclohexadienone *98* [117]. Märkl has shown that the reaction takes place *via* a series of As-alkylations followed by facile heterocope rearrangements (→*99*→*100*→etc.). The analogous reaction with propargyl bromide yields arsachromone *102* [118].

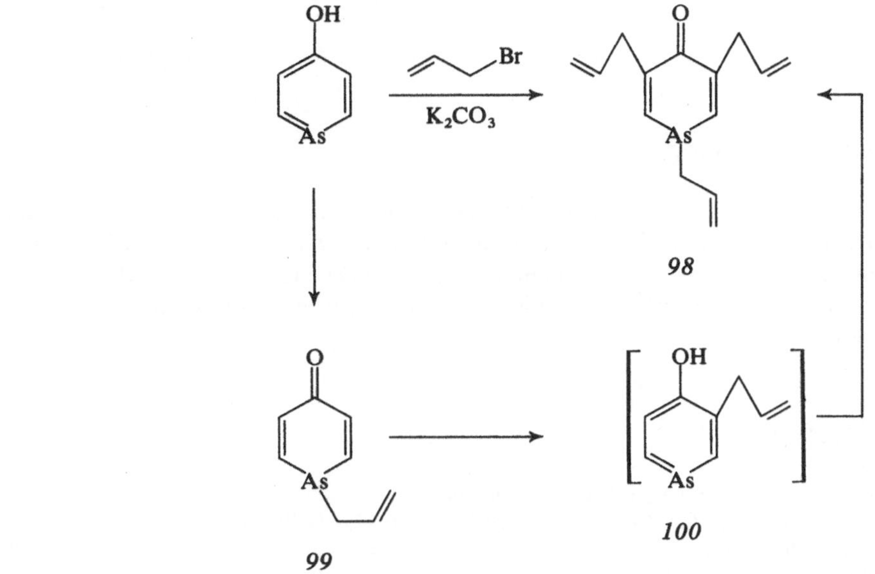

All three arsabenzene carboxylic acids are available [119,120]. pK$_a$ measurements indicate that they are all stronger acids than benzoic acid. This seems to imply that the carbon atoms of arsabenzene are slightly electron deficient relative to benzene.

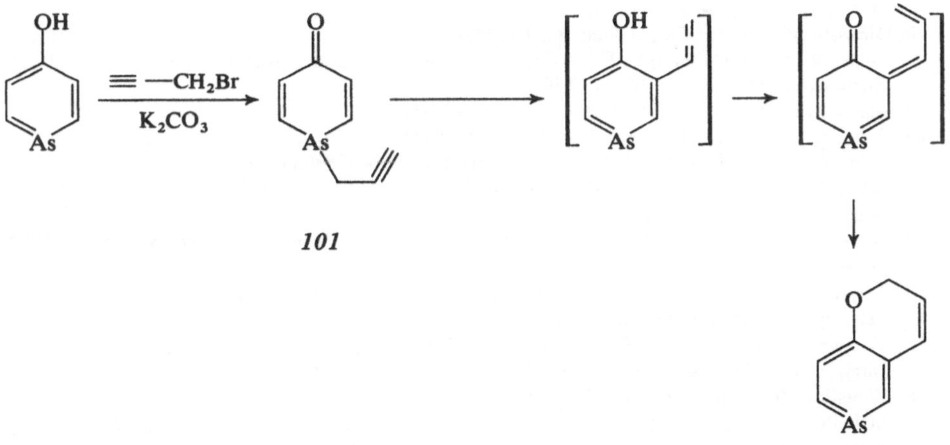

151

Arthur James Ashe

MO calculations have indicated that the 3-position of arsabenzene is the most electropositive [66, 69]. The observation that 3-arsabenzoic acid is the strongest carboxylic acid is consistent with this.

$pK_a = 4.2$ $pK_a = 3.8$ $pK_a = 3.7$ $pK_a = 4.1$

5 Acknowledgments

This work was partially supported by the National Science Foundation and the National Institutes of Health. It is a pleasure to acknowledge the fine work on arsabenzene done by Professors Bickelhaupt and Märkl and their coworkers. I am personally grateful to my own able coworkers whose accomplishments I have described.

6 References

1. Pitzer, K. S.: J. Am. Chem. Soc. 70, 2140 (1948)
2. Mulliken, R. S.: J. Am. Chem. Soc. 72, 4493 (1950); 77, 884 (1955)
3. Douglas, B. E., McDaniel, D. H.: Concepts and Models of Inorganic Chemistry, Blaisdell Publ. Co., New York, N.Y., 1965, p. 59
4. Märkl, G.: Angew. Chem. Int. Ed. Engl. 5, 846 (1966)
5. Märkl, G.: Lect. Heterocyclic Chem. 1, S-69 (1972)
6. Märkl, G.: Phosphorus and Sulfur 3, 77 (1977)
7. Dimroth, K.: Topics Curr. Chem. 38, 1 (1973)
8. Jongsma, C., Bickelhaupt, F.: Topics in Non-Benzenoid Aromatic Chemistry, vol. II, Hirokowa Publ., Tokyo, 1977, p. 139
9. Ashe, A. J., III: J. Am. Chem. Soc. 93, 3293 (1971)
10. Ashe, A. J., III: J. Am. Chem. Soc. 93, 6690 (1971)
11. Ashe, A. J., III, Gordon, M. D.: J. Am. Chem. Soc. 94, 7596 (1972)
12. Jutzi, P.: Angew. Chem. Int. Ed. 14, 232 (1975)
13. Ashe, A. J., III: Acc. Chem. Res. 11, 153 (1978)
14. Quin, L. D.: The Heterocyclic Chemistry of Phosphorus, Wiley-Interscience, New York, 1981, pp. 141–152, 390–401
15. Tzschach, A., Heinicke, J.: Arsenheterocyclen, VEB Deutscher Verlag für Grundstoffindustrie, Leipzig, 1978, pp. 124–130, 135–138
16. Vermeer, H., Bickelhaupt, F.: Angew. Chem. Int. Ed. 8, 992 (1969)
17. Jutzi, P., Deuchert, K.: ibid. 8, 991 (1969)
18. Weustink, R. J. M., Jongsma, C., Bickelhaupt, F.: Tetrahedron Lett. 1975, 199
19. Bickelhaupt, F., Lourens, R., Vermeer, H.: Rec. Trav. Chim. Pays-Bas 98, 3 (1979)
20. Vermeer, H., Bickelhaupt, F.: Tetrahedron Lett. 3255 (1970)
21. deKoe, P., Bickelhaupt, F.: Angew. Chem. Int. Ed. 6, 567 (1967)
22. deKoe, P., Bickelhaupt, F.: ibid 7, 889 (1968)
23. Ashe, A. J., III, Shu, P.: J. Am. Chem. Soc. 93, 1804 (1971)
24. Ashe, A. J., III: Tetrahedron Lett. 1976, 415
25. Ashe, A. J., III, Chan, W.-T., Perozzi, E.: ibid. 1975, 1083

26. Ashe, A. J., III, Chan, W.-T.: J. Org. Chem. *44*, 1409 (1979)
27. Ashe, A. J., III, Chan, W.-T.: Tetrahedron Lett. *1975*, 2749
28. Märkl, G., Baier, H., Heinrich, S.: Angew. Chem. Int. Ed. *14*, 710 (1975)
29. However, there are limitation of substituents: *26c* (R = OSi(CH$_3$)$_3$ leads to very poor yields of *27c*. See: Märkl, G., Hofmeister, P., Kneidl, F.: Tetrahedron Lett. *1976*, 3125
30. Jutzi, P., Baumgärtner, J.: J. Organometal. Chem. *148*, 247 (1978)
31. Smith, T. W.: Ph.D. Thesis, University of Michigan, 1977
32. El-Sheikh, M. Y.: unpublished
33. Diephouse, T. R.: Ph.D. Thesis, University of Michigan, 1981
34. Märkl, G., Kneidl, F.: Angew. Chem. Int. Ed. *12*, 931 (1973)
35. 3-Substituted pentadiyn-3-ols may also be converted to 4-substituted arsabenzenes, although the reactions are more complicated. See: Märkl, G., Rampal, J. B.: Tetrahedron Lett. *1977*, 2325
36. Märkl, G., Kneidl, F.: Angew. Chem. Int. Ed. *13*, 667 (1974)
37. Märkl, G., Kneidl, F.: ibid. *13*, 668 (1974)
38. Märkl, G., Kellerer, H., Kneidl, F.: Tetrahedron Lett. *1975*, 2411
39. Märkl, G., Kellerer, H.: Tetrahedron Lett. *1976*, 665
40. Märkl, G., Advena, J., Hauptmann, H.: ibid. *1974*, 203
41. Märkl, G., Advena, J., Hauptmann, H.: Angew. Chem. Int. Ed. *11*, 441 (1972)
42. Märkl, G., Advena, J., Hauptmann, H.: Tetrahedron Lett. *1974*, 303
43. Märkl, G., Hauptmann, H.: ibid. *1968*, 3257
44. Sanz, F., Daly, J. J.: Angew. Chem. Int. Ed. *11*, 630 (1972); J. Chem. Soc., Dalton Trans. *1973*, 511
45. Lattimer, R. P., Kuczkowski, R. L., Ashe, A. J., III, Meinzer, A. L.: J. Mol. Spectrosc. *57*, 428 (1975)
46. Fong, G., Kuczkowski, R. L., Ashe, A. J., III: ibid. *70*, 197 (1978)
47. Wong, T. C., Ashe, A. J., III, Bartell, L. S.: J. Mol. Struct. *25*, 65 (1975)
48. Wong, T. C., Bartell, L. S.: ibid. *44*, 169 (1978)
49. Wong, T. C., Ashe, A. J., III: ibid. *48*, 219 (1978)
50. Wong, T. C., Ferguson, M. G., Ashe, A. J., III: ibid. *52*, 231 (1979)
51. Cruickshank, D. W. J., Sparks, R. A.: Proc. Roy. Soc. London, Ser. A *258*, 270 (1960); Almenningen, A., Bastiansen, O., Dyvik, F.: Acta Crystallogr. *14*, 1056 (1961); Pawley, G. S., Yeats, E. A.: Acta Crystallogr., Sect. B *25*, 2009 (1969)
52. Sorensen, G. O., Mahler, L., Rastrup-Andersen, N.: J. Mol. Struct. *20*, 119 (1974)
53. Gibbs, J. H.: J. Chem. Phys. *22*, 1460 (1954)
54. Ashe, A. J., III, Sharp, R. R., Tolan, J. W.: J. Am. Chem. Soc. *98*, 5451 (1976)
55. Dorfman, Ya. G.: Diamagnetism and the Chemical Bond, Arnold, London, 1965, Chapt. 2.
56. McConnell, H. M.: J. Chem. Phys. *27*, 226 (1957)
57. Electronegativity effects were corrected using the equation $\Delta(\delta) = 1.16(N_E - N_C)$ where N is the Allred-Rochow electronegativity. See: Narasimhan, P. T., Rogers, M. T.: J. Am. Chem. Soc. *82*, 5983 (1960)
58. Karplus, M., Pople, J. A.: J. Chem. Phys. *38*, 2803 (1963)
59. Shoolery, J. N.: ibid. *31*, 1427 (1959)
60. Batich, C. et al.: J. Am. Chem. Soc. *95*, 928 (1973)
61. Bastide, J. et al.: Tetrahedron Lett. *1976*, 411
62. Ashe, A. J., III et al.: Helv. Chim. Acta *59*, 1944 (1976)
63. Burrow, P. D., Ashe, A. J., III, Bellville, D. J., Jordan, K. D.: J. Am. Chem. Soc. submitted (1981)
64. For a general discussion of the application of PE spectra to these and similar compounds see: Bock, H.: Pure Appl. Chem. *44*, 343 (1975); Heilbronner, E., Maier, J. P., Haselbach, E.: Phys. Meth. Heterocycl. Chem. *6*, 1 (1974)
65. Jordan, K. D., Burrow, P. D.: Acc. Chem. Res. *11*, 341 (1978)
66. Oehling, H., Schäfer, W., Schweig, A.: Angew. Chem. Int. Ed. *10*, 656 (1971)
67. von Niessen, W., Diercksen, G. H. F., Cederbaum, L. S.: Chem. Phys. *10*, 345 (1975)
68. Palmer, M. H. et al.: J. Chem. Soc. Perkin Trans. 2 *1975*, 841
69. Clark, D. T., Scanlan, I. W.: J. Chem. Soc., Faraday Trans. 2 *70*, 1222 (1974)
70. Hase, H. L., Schweig, A., Hahn, H., Radloff, J.: Tetrahedron *29*, 475 (1973)

71. Rajzmann, M., Francois, P., Carles, P.: J. Chim. Phys. *76*, 328 (1979)
72. Herndon, W. C.: Tetrahedron Lett. *1979*, 3283
73. Ilić, P., Sinković, B., Trinajstić, N.: Isr. J. Chem. *20*, 258 (1980)
74. Fukui, K. et al.: J. Chem. Phys. *22*, 1433 (1954)
75. Jongsma, C. et al.: Tetrahedron *31*, 2931 (1975)
76. Gibbs, J. H.: J. Phys. Chem. *59*, 644 (1955)
77. Kuz'min, K. I., Kamai, G.: Dokl. Acad. Nauk SSSR *73*, 709 (1950)
78. Aroney, M. J., Le Fèvre, R. J. W., Saxby, J. D.: J. Chem. Soc. *1963*, 1739
79. Davies, W. C.: ibid *1935*, 462
80. Liu, R. S. H.: J. Am. Chem. Soc. *90*, 215 (1968)
81. Streitwieser, A., Jr.: Molecular Orbital Theory for Organic Chemists, John Wiley and Sons, New York, 1961, pp. 432–438
82. Doak, G. O., Freedman, L. D.: Organometallic Compounds of Arsenic, Antimony and Bismuth, John Wiley and Sons, New York, N.Y., 1970, p. 8
83. Reliable values for E-E bond energies are not available. However, the strength of a C-C bond (82.5 kcal/mol) exceeds that of twice a Bi-C bond (34.1 kcal/mol) by 15 kcal/mol. On this basis, dimer *45b* is estimated to be at least 15 kcal/mol more stable than dimer *46b*. For bond energies see: Skinner, H.A.: Adv. Organometal. Chem. *2*, 49 (1964)
84. Kreil, C. L. et al.: J. Am. Chem. Soc. *102*, 841 (1980)
85. However, 1,4-di-t-butylsilabenzene forms a 1,2-2,1-dimer. See: Markl, G., Hofmeister, P.: Angew. Chem. Int. Ed. *18*, 789 (1979)
86. Märkl, G., Advena, J., Hauptmann, H.: Tetrahedron Lett. *1972*, 3961
87. Ashe, A. J., III, Friedman, H. S.: ibid. *1977*, 1283
88. Ashe, A. J., III, Chan, W.-T., Smith, T. W.: ibid. *1978*, 2537
89. Ashe, A. J., III et al.: J. Org. Chem. *46*, 881 (1981)
90. Ashe, A. J., III, Bellville, D. J., Friedman, H. S.: J. Chem. Soc., Chem. Commun. *1979*, 880
91. Smith, T. W.: Ph.D. Thesis, University of Michigan, 1977
92. Colburn, J. C.: Ph.D. Thesis, University of Michigan, 1978
93. Kolling, O. W., Mawdsley, E. A.: Inorg. Chem. *9*, 408 (1970)
94. Hodges, R. V., Beauchamp, J. L., Chan, W.-T., Ashe, A. J., III: J. Am. Chem. Soc., *1981*, in press
95. For trimethylphosphine < CPC = 99°: Bryan, P. S., Kuczkowski, R. L.: J. Chem. Phys. *55*, 3049 (1971). For trimethylarsine < CAsC = 96°: Lide, D. R., Jr.: Spectrochim. Acta *15*, 473 (1959)
96. Staley, R. H., Beauchamp, J. L.: J. Am. Chem. Soc. *96*, 6252 (1974)
97. The bond angles for NH_3 and the methylamines vary from 100–112°, which are rather close to the 109.5° expected for the corresponding ammonium salts. For structural data on nitrogen compounds see: For NH_3: Coles, D. K. et al.: Phys. Rev. *82*, 877 (1951); for CH_3NH_2: Lide, D. R., Jr: J. Chem. Phys. *27*, 343 (1957); for $(CH_3)_3N$: Lide, D. R., Jr., Mann, D. E.: ibid. *28*, 572 (1958)
98. Ashe, A. J., III et al.: J. Am. Chem. Soc. *101*, 1764 (1969)
99. Martin, R. L., Shirley, D. A.: J. Am. Chem. Soc. *96*, 5299 (1974); Davis, D. W., Rabalais, J. W.: ibid, *96*, 5305 (1974); Carroll, T. X., Smith, S. R., Thomas, T. D.: ibid. *97*, 659 (1978); Mills, B. E., Martin, R. L., Shirley, D. A.; ibid. *98*, 2380 (1970); Smith, S. R., Thomas, T. D.: ibid. *100*, 5459 (1978)
100. Ashe, A. J., III, Colburn, J. C.: J. Am. Chem. Soc. *99*, 8099 (1977)
101. It might be noted that the ⋆ CAsC of trimethylarsine expanded to 106° in $(Me_3As)_2Pt_2Cl_4$. The difficulty of achieving this change for arsabenzene might explain its weakness as a ligand. See: Watkins, S. F.: J. Chem. Soc. A *1970*, 168
102. Märkl, G., Liebl, R.: Angew. Chem. Int. Ed. *16*, 637 (1977)
103. Märkl, G., Rampal, J. B.: Tetrahedron Lett. *1977*, 3449; *1978*, 1471
104. Märkl, G. Rampal, J. B.: ibid. *1978*, 1175
105. Boar, R. B. et al.: J. Chem. Soc., Perkin I *1975*, 1237
106. Märkl, G., Rampal, J. B.: Tetrahedron Lett. *1977*, 2569
107. Chan, W.-T.: Ph.D. Thesis, The University of Michigan, 1977
108. See: Costin, C. R., Morrow, C. J., Rappoport, H.: J. Org. Chem. *41*, 535 (1976)

109. Wiberg, K. B., Kintner, R. R., Motell, E. T.: J. Am. Chem. Soc. *85*, 450 (1963)
110. Weustink, R. J. M., Jongsma, C., Bickelhaupt, F.: Rec. Trav. Chim. Pays-Bas. *96*, 265 (1977)
111. Weustink, R. J. M., Lourens, R., Bickelhaupt, F.: Liebigs Ann. Chem. *1978*, 214
112. Ziegler, K., Zeiser, H.: Ber. dtsch. chem. Ges. *63*, 1847 (1930); Liebigs Ann. Chem. *485*, 174 (1931)
113. Ashe, A. J., III, Smith, T. W.: Tetrahedron Lett. *1977*, 407
114. Ashe, A. J., III, Smith, T. W.: J. Am. Chem. Soc. *98*, 7861 (1976)
115. Märkl, G., Rampal, J. B., Schöberl, V.: Tetrahedron Lett. *1979*, 3141
116. Märkl, G., Rampal, J. B.: ibid. *1976*, 4143
117. Märkl, G., Rampal, J. B.: Angew. Chem. Int. Ed. *15*, 690 (1976)
118. Märkl, G., Rampal, J. B.: Tetrahedron Lett. *1979*, 1369
119. Märkl, G., Kellerer, H.: ibid. *1976*, 665
120. Ashe. A. J., III, Chan, W.-T.: J. Org. Chem. *45*, 2016 (1980)

Author Index Volumes 101–105

Contents of Vols. 50–100 see Vol. 100
Author and Subject Index Vols. 26–50 see Vol. 50

The volume numbers are printed in italics

Recktenwald, O., see Veith, M.: *104*, 1–55 (1982).
Rolla, R., see Montanari, F.: *101*, 111–145 (1982).
Rzaev, Z. M. O.: Coordination Effects in Formation and Cross-Linking Reactions of Organotin Macromolecules. *104*, 107–136 (1982).

Saenger, W., see Hilgenfeld, R.: *101*, 3–82 (1982).
Steudel, R.: Homocyclic Sulfur Molecules. *102*, 149–176 (1982).
Steudel, R., and Laitinen, R.: Cyclic Selenium Sulfides. *102*, 177–197 (1982).

Veith, M., and Recktenwald, O.: Structure and Reactivity of Monomeric, Molecular Tin(II) Compounds. *104*, 1–55 (1982).
Voronkov, M. G., and Lavrent'yev, V. I.: Polyhedral Oligosilsequioxanes and Their Homo Derivatives. *102*, 199–236 (1982).

J. Fabian, H. Hartmann

Light Absorption of Organic Colorants

Theoretical Treatment and Empirical Rules

1980. 76 figures, 68 tables. VIII, 245 pages
(Reactivity and Structure, Volume 12)
ISBN 3-540-09914-X

Contents: Phenomenological Conceptions on Color
and Constitution. – UV/VIS Spectroscopy and
Quantum Chemistry of Organic Colorants. – Relation
Between Phenomenological and Quantum Chemical
Theories. – Theoretical Methods for Deriving Color-
Structure Relationships – Classification of Organic
Colorants. – Polyene Dyes. – Azo Dyes. – Carboxi-
mide, Nitro and Quinacridone Dyes. – Quinoid
Dyes. – Indigoid Dyes. – Diphenylmethane, Triphenyl-
methane and Related Dyes. – Polymethine Dyes. –
Porphyrins and Phthalocyanines. – Conjugated Betaine
Dyes. – Multiple Chromophore Dyes. – References. –
Subject Index.

H. Primas

Chemistry, Quantum Mechanics and Reductionism

Perspectives in Theoretical Chemistry

1981. XII, 451 pages
(Lecture Notes in Chemistry, Volume 24)
ISBN 3-540-10696-0

Contents: Open Problems of the Present-Day Theoretical Chemistry. – On the Structure of Scientific
Theories. – Pioneer Quantum Mechanics and its Inter-
pretation. – Beyond Pioneer Quantum Mechanics. –
A Framework for Theoretical Chemistry. – Reduc-
tionism, Holism and Complementarity. – Bibliography
and Author Index. – Index.

Springer-Verlag
Berlin
Heidelberg
New York

M. F. O'Dwyer, J. E. Kent, R. D. Brown

Valency

Heidelberg Science Library

2nd edition. 1978. 150 figures. XI, 251 pages
ISBN 3-540-90268-6

Contents: Gross Atomic Structure. – Atomic Theory. – Many-Electron Atoms. – Molecular Theory and Chemical Bonds. – The Solid State. – Experimental Methods of Valency.

This textbook is designed for use by advanced first year freshman chemistry students as well as physical chemistry students in their sophomore and junior years.

It covers SI units and the concept of energy, and the structure and theory of atoms, using wave mechanics and graphs to define atomic orbitals and the meaning of quantum numbers, for both hydrogen atoms as well as many-electron atoms. Periodic trends such as ionization and orbital energies are emphasized and explained through atomic theory.

The book also covers molecular theory and the chemical bond using a model approach. Electrostatic models for ionic compounds and transition metal complexes and a molecular orbital are included together with valence-bound and Sidgwick-Powell models for covalent compounds. Problems and appendices are provided to enable readers to deepen their comprehension of the subject.

Springer-Verlag
Berlin
Heidelberg
New York